Horses

THEIR SELECTION, CARE AND HANDLING

BY

MARGARET CABELL SELF

Author of
TEACHING THE YOUNG TO RIDE

1976 EDITION

Published by
Melvin Powers
WILSHIRE BOOK COMPANY
12015 Sherman Road
No. Hollywood, California 91605
Telephone: (213) 875-1711

A. S. Barnes and Co., Inc.
Cranbury, New Jersey

Thomas Yoseloff, Ltd
18 Charing Cross Road
London W. C. 2, England

Twenty-first Printing September, 1967

Printed by

HAL LEIGHTON PRINTING CO.
P.O. Box 1231
Beverly Hills, California 90213
Telephone: (213) 983-1105

08302
PRINTED IN THE UNITED STATES OF AMERICA
ISBN 0-87980-195-6

Preface

"There is something about the outside of a horse that is good for the inside of a man" is a saying that has been familiar to horsemen for a good many years. But this does not mean merely owning a horse and riding or driving him. It means the care of him as well, the grooming, the shovelling and everything else. When doctors of the nineteenth century advised mothers to get ponies for their not-too-husky children, or suggested to liverish men suffering from an over-abundance of good, red meat, that they invest in a horse and look after it themselves, they knew that the exercise entailed was just the kind of exercise needed by their patients, and that the mental effect of developing such an absorbing hobby would also be all to the good. It is certain that the man, woman or child who owns, loves and cares for one particular animal, derives ten times more pleasure from his possession than the one who merely hires a hack when he wants to ride; or, owning a horse, boards him out. The hours of physical labour involved in caring for your mount, the hours of solving the problems that come up and of devising means of curing this habit or improving that, will show a heavy yield in both health and enjoyment.

But beware of one thing—to a great extent, animals mirror the dispositions and characters of their owners! The master may seek in every way possible to develop and train his dog or horse to be gentle, brave, steady and courageous, but if he himself is nervous, timid, erratic or mean the animal will, after a comparatively short association, begin to show these same traits! In the dog world one sees litter mates, parcelled out to different families, turn out as varied in character as though they were entirely unrelated, and not from any particular or intentional training on the parts of their various masters. One pup may go to a family that is friendly and casual, another to a family of high-strung, nervous people, and a third to a family more interested in dogs as ornaments than as companions. So the first becomes the sort that greets you with a wag of the tail, the second approaches apparently expecting the worst and consequently is on the defensive, just in case, while the third spends most of his time away from home preying on the neighbour's chickens. In the same way, if you go into a stable or walk into a field of horses, you know immediately something of the character of the man or woman who looks after them.

Remember that a horse is nervous and timid by nature. He is not vicious but he may be willful. He is not intelligent, in that he can think of only one thing at a time and cannot reason, but he can be crafty. It is the rider and the trainer who must do the thinking. There is no reason for fear if he thinks clearly and does not put himself into a situation whereby his own, ill-considered actions will cause undesirable reactions on the part of the horse.

In this book my first aim has been to describe the more familiar breeds of horses to be found in this country, their characteristics and the purposes to which they may be put; also the various kinds of saddles, bridles, bits, vehicles, etc. from which one may choose, as well as the essential stable accessories. I have written at some length on the everyday handling of horses as apart from their care and training. These things may seem obvious to the man or woman who has always looked after his own animal but they are points which often do not occur to the novice, or to some one who, having perhaps ridden a great deal, yet has never actually worked around horses. They are the "tricks of the trade" which make things easier.

I have gone into the feeding, grooming and general care of the horse, also the cost in hours and money so that a person planning to own a horse or pony for the first time will be able to judge what this entails. I have discussed common stable vices and how to handle them as well as other vices. The chapter on riding is not intended to be more than a brief summary of the modern ideas of horsemanship. I hope the chapter on teaching children will be useful to parents when they decide to introduce their offspring to the science of equitation. Driving is coming back into fashion and I have emphasised the importance of well-fitting harness and suitable vehicles as well as the more common "do's and don'ts." The chapter on First

Aid is not intended to replace the veterinary. It is simply what it implies, the immediate care of the ordinary run of minor illnesses and accidents that occur in any stable.

I should like to express my appreciation to the Smith-Worthington Saddlery Company for supplying most of the illustrations of saddle and bridle equipment.

I trust this book will be found of value to any one who loves horses to the extent of wanting to have and care for his own, and that it will encourage such a person to try the fun and satisfaction of being his own groom and stable man.

MARGARET CABELL SELF

(Photo by Anthony Lanza)

Thoroughbred Brood Mares with Their Foals

Contents

Illustrations

HORSES

Their Selection, Care and Handling

CHAPTER I

Selection of the Horse

The original horse was a little animal, the size of a fox, with soft toes and no defence from enemies except his ability to outrun them. Later, after centuries of such running during which he had attained size, speed and a hard hoof, man decided to utilise the horse for the purposes of warfare. As time went on, various types of horses were developed according to other specific needs. Thus the Roman horses were stocky with heavy shoulders and necks for pulling weight. The Romans did not understand the principles of drawing a load—their horses actually *pulled* the chariots, whereas in modern harness the horse does not *pull* at all, he *pushes* against the collar and is thus able to draw more than his own weight. The Romans used three horses to pull a light chariot and one man! In the Middle Ages a very heavy type of horse was developed in Europe, this type, the forerunner of the modern draught horse, was needed to carry the tremendous weight of a knight in armour.

In Arabia men needed a swift steed that could outrun the enemy, could be used on soft ground, and would not need to carry much weight. As a result the ARABIAN horse of to-day is speedy, light of bone, sensitive and nervous. He is gentle but his reactions are so quick that an indifferent rider sometimes has trouble retaining his seat. The Arabs loved their horses and made companions of them, consequently Arabian horses are more affectionate and more highly developed in intelligence than some of the other breeds.

Modern times, with varied demands, have produced many different established and recognised breeds of horses. The THOROUGH-BRED, descended from Arabian stallions bred to European mares, was developed for speed and courage. He is more sensitive but less steady and hardy than some other breeds. In conformation he carries his head out rather than up and does not lift his feet far off the ground, thus his stride is longer and he can cover more ground in

A Splendid, Heavy Weight Thoroughbred Type

Notice the fineness of his head, a characteristic inherited from the Arabian strain. On page 160 you will see a picture of him taking the hog's back.

less time than the high stepping saddlers. He has a tremendous heart and has been known to die on the race track rather than give up. He is not suited to draught purposes though he can pull a light weight, especially if he is the short-coupled type rather than the rangy type. Most thoroughbreds, because of the sensitiveness and quickness of reaction inherited from the Arab, make poor mounts for beginners. However, a mixture of thoroughbred blood does much to improve other breeds for general saddle and harness purposes as it adds speed and willingness.

It might be well here to differentiate between the noun "thorough-bred" and the adjective "thoroughbred." The noun refers to a specific breed of animal, as one speaks of a collie or a setter meaning an established breed of dog. However, many people use it to mean "purebred." They will ask if a five-gaited Kentucky saddle horse is a "thoroughbred," meaning if it is of registered ancestry. This is an incorrect usage of the word and should be avoided.

By its practice of placing thoroughbred stallions out to service the Remount Division of the United States Cavalry has done much to improve the blood of farm and ranch owned stock throughout the country. The farmer who keeps the stallion collects a very moderate stud fee from the owners of the neighbouring cold blooded mares. Four years later the Army sends its men to buy up likely looking colts. In this way a large percentage of cavalry horses have the desirable strain of thoroughbred blood in them.

THE KENTUCKY SADDLE HORSE, a breed diametrically opposed to the thoroughbred in appearance, and bred for very different purposes was developed on the southern plantations. The plantation owner needed a horse with an easy gait that would carry him comfortably all day while he rode over his fields. For show purposes the animal has been bred for high action, spirit, beauty and quality, but not for speed. The Kentucky saddler has five gaits, showing the "slow gait," a shuffling, four beat gait that is smooth and not tiring and the fast "rack" or "singlefoot," also a smooth, four beat gait, in addition to the regulation walk, trot and canter. The saddler has a broad, showy front and short back. He holds his head high and picks up his feet until his knees almost touch his chin. He wears his mane and fetlocks long, his tail full. He may also be used in harness. The three gaited saddler is lighter in build than his five-gaited cousin and shows only the walk, trot and collected canter. His action, too, is showy, he wears a clipped mane and he too may be used as a combination "ride-and-drive."

The highly bred examples of these breeds are gentle and loveable, but they are too sensitive and spirited for the average rider. In re-

cent years a new breed has become established known as the PLAN-
TATION WALKER. These horses must show the slow gait and a straight
canter (they must not travel on a diagonal at the canter) so that the
plantation owner can ride between his corn and cotton rows without treading on the crops. Horses from common mares bred to high
class saddle stallions usually make good-looking, good-tempered
combination horses with easy gaits. They are tractable, easily
handled in the stable and the extreme sensitivity of the saddler is
tempered by the more phlegmatic disposition of the less highly bred
animal.

THE MORGAN HORSE is a sturdy, small breed of horse descended
from the famous stallion Justin Morgan. Originally a driving and
light work breed, they have tremendous fronts and shoulders,
rather long backs, light quarters and short legs. They are not excitable and make splendid light-weight harness horses or saddlers,
especially for children. The Morgan Horse Association holds a hundred mile ride in Vermont every year to encourage the development
of the Morgan horse. Under test conditions this little horse has
proved again and again its great staying powers, tractability and
hardiness. There is no more satisfactory pony for the small beginner
than the cross between either the Welsh or Shetland pony and the
Morgan.

Curiously enough, the standard, or trotting breeds, of which the
HAMBLETONIAN is an example, came to be because of the Puritan
Blue Laws against horse racing. It was illegal to ride a race, but
there was nothing on the books to say that one farmer, meeting another in his cutter, couldn't have a friendly "brush" down the main
street of the village. The modern trotter, put to a featherweight
sulky with bicycle wheels, covers a mile in under two minutes. He is
long barrelled and high rumped with plenty of heart. He is tractable
and kindly, not nearly so jittery as the Thoroughbred or the Arab,
and shows incredible speed at the trot. The canter is not natural to
him and it takes a great deal of patience to teach the trotter a show
canter. Those trained for the track are not suitable for other purposes. They are schooled to a tight rein, the harder you pull, the
faster they go, which can be very disconcerting to the amateur
driver! But a colt of trotting blood trained to the saddle and to drive
makes an excellent combination horse and some of them take readily
to jumping as well. These horses are gentle yet spirited and they

Characteristic Scene on the Western Plains

There are several thousand horses in this herd that is being moved toward the home ranch so that they can be looked over, the foals branded for identification, and the colts and fillies of three or four years broken to the saddle preparatory to being sold. Notice the dust that is kicked up by the thousands of hooves and the wrangler who stands ready to herd back any stray animal.

seldom have any vices. They are friendly and they will get you where you want to go fast!

The WESTERN HORSE or cowpony is descended from the Spanish horses imported at the time of the Spanish invasion of Mexico. The horse was not native to North America. The cowpony is not a standardised breed but has been developed on the plains of this country for use in the herding of cattle. They are bred and trained for speed at the gallop, quickness in turning and flexibility, extreme hardiness and stamina. These ponies, crossed with the thoroughbred, provide the bulk of America's polo ponies and many of this cross are also used by the Army. The western horse, partly because of the way he is broken, is apt to be trickier than the other breeds. He will buck more readily and is harder to handle and less gentle. He does not seem to have much affection for man, but he will keep in good condition and take a lot of hard work on forage that would mean starvation to the more highly bred animal. Nor is he as much bothered by flies. Turn a bunch of horses, some Thoroughbreds and some Westerns out to pasture that is not too luxuriant, during the fly season, and when you bring them in the Westerners will be fat and frisky while the Thoroughbreds will look about ready for the bone yard.

The WESTERN CHUNK is a light draught breed developed in the west and middle west. These horses run around twelve to fifteen hundred pounds in weight although they are not usually more than fifteen hands in height. They are phlegmatic, gentle, easily trained and slow. They make an ideal mount for the heavy weight man who wants mild exercise and no excitement. Timid drivers may be sure of getting to their destination although it will take a little time. The "chunk" will also earn his keep doing work around the farm. If he is to be used under the saddle as well as between the shafts, select a horse with not too broad a back. See page 38.

The heavy draught horse, BELGIAN, PERCHERON, SHIRE, etc. is phlegmatic in the extreme and very gentle. Because of his great size he needs a lot of hay to keep him going but not a proportionate increase in grain inasmuch as his work is not fast. These animals sometimes attain a weight of twenty-three hundred pounds and eighteen or nineteen hands in height. Crossed with the Thoroughbred they make excellent heavy weight hunters, particularly if the draught mare is bred to the Thoroughbred stallion and the resulting foal bred to another Thoroughbred.

Before going on to discuss the common breeds and types of ponies it might be well to explain that in horseman's parlance the term "pony" denotes any animal of the equine race that is under fourteen hands and two inches (fifty eight inches) in height. The measure is taken at the withers. Thus a so-called "pony" can be part or all horse as far as ancestry goes and many of the best ones are. The ponies of the English Isles that just grew without interference from

A Pair of Heavy Western Chunks

They make fine all-purpose horses, especially those not quite as big as the ones shown. They are phlegmatic and amiable

mankind, adapted themselves to the natural conditions. They became hardy, small and wise. Their feet are tougher than those of horses, they can subsist on very little feed and they are independent and somewhat stubborn with a lot of personality. The Mongolian pony, a slightly larger breed, that developed on the rough cold wastes of the Mongolian desert is ugly, sturdy and swift. In their native wilderness they keep fit on very little forage. When brought in, groomed and fed they seem to lose much of their speed as well as their hardiness. These ponies were the original polo ponies, and in China to-day it is not uncommon to see a six foot man mounted on a thirteen hand pony, his feet nearly dragging on the ground. There is nothing more satisfactory for the very young rider than a well-mannered, well-trained, active pony, but they are very hard to find. One to be used for riding should not be too broad of beam nor have too thick a neck. He should be of a size and nature suitable to the child who is going to ride him. The beginner will want the quietest pony obtainable, but the child who is a good rider will get an enormous amount of pleasure out of a spunky, fast, flexible pony that will carry him cross-country at a gallop and on whom he can be the Lone Ranger in person. At the same time the inevitable tumbles will mean little because of the short distance involved. If you have two children get two ponies—they work better in pairs.

The English Type of Shetland Pony

Contrast with the picture of the American Shetland on page 9. The Rider here demonstrates an excellent seat without saddle, body straight, legs relaxed.

The ENGLISH SHETLAND, built like a miniature Percheron is not suitable for riding because of his conformation, but he is tireless in the cart and can draw heavy loads.

The AMERICAN SHETLAND has been bred to a slimmer, finer type, also larger. He is smart as a whip and if you find a really good one he will be both companion and servant to your child. Your only difficulty will be in figuring out a way to keep him from getting out.

American Shetland Type

This is a more slender animal than the English Shetland shown on page 8. He will carry his young rider across country and over the jumps following the bloodhound which you see at his feet.

Grass on the other side of the fence is always the greenest to a pony and he can soon figure out how to open gates and take down draw-bars. Don't feed either of these breeds grain unless they are in very poor condition or having more than two hours work a day. Ponies fed grain develop the bad habit of nipping and soon become too fresh for their young riders.

The HACKNEY PONY is a harness type with high, showy action. He is nervous and sensitive and the highly bred ones are usually only suitable for show purposes, but Hackneys crossed with other breeds make fine ponies for children either for riding or driving or both. Watch out for roughness of gait at the trot on the ones with exaggerated action. It is hard for a beginner to become accustomed to this. Hackneys may be anywhere from eleven to fourteen hands in height. Those crossed with Welsh strains sometimes make good hunting and jumping ponies. They are not as stubborn as the Shet-lands.

The so-called "INDIAN PONY" is not a recognised breed, rather it is a small edition of the western cowpony though sometimes stockier. This type is tireless, quite tractable and, if properly broken, makes a good mount for an active youngster.

The WELSH PONY is built like a little horse and is an excellent breed for children, he is a natural jumper, has a good disposition and is rugged.

In addition to these established breeds one runs across in-numerable ponies of no recognised ancestry that are sometimes the most useful of all except for show purposes, having the hardiness and stamina of the pony with the gentleness of the horse.

It can easily be seen from the foregoing that before setting out to buy a horse one should first consider very seriously exactly what kind of work the horse is going to be required to do as well as the general ability, disposition and preference of the prospective owner. The experienced, sporting type of man or woman who wants a high stepper, an open jumper or a fast trotter would never be satisfied with the general, all around useful beast that would be priceless to the family with children who wanted something safe and sure. Every horse you buy is a compromise. You may set out with the purpose of purchasing a certain type, complete in detail in your mind down to the color, and you are very apt to return with some-thing quite different but just as useful.

WHERE SHOULD ONE LOOK FOR A HORSE?

The ideal thing to do, if you want one of the established breeds, is to go straight to the BREEDING FARM that raises this type. A letter to the Horse and Mule Association will give you the names of breeders in your locality. A quick perusal of the horse advertisements in the sporting magazines and in the sports sections of the papers will tell you of others. By going to the breeder you will have a good selection from which to choose. Sometimes a personal visit is not necessary, you can write and describe exactly the kind of animal you want and the price you want to pay, and the breeder will tell you honestly whether or not he thinks he can suit you. For their own sakes breeders will do their best to please you knowing that their reputations rest on the number of customers they are able to satisfy. Often a breeder will sell a horse on the understanding that if he does not come up to expectations or is not as guaranteed as you may choose another.

If you have no standard breed in mind the next best place to go is to the DEALER. This is a man who runs a sales stable, having a number of animals at all times, some of which he owns and some of which he may have on commission. He also probably knows of all the horses for sale within vanning distance. Here you run somewhat more of a chance of not getting your money's worth than you do at a breeders. The dealer himself may not know too much about the manners and dispositions of the horses that may have been with him for only a very short space of time. However, if you have it understood that if the horse is unsatisfactory you can bring him back; and especially if you can get a veterinary's certificate of soundness, you will probably run into no trouble. See the animal several times and ride him or drive him under as many conditions as possible. Get in touch with the former owner if you can and ask all the questions you can think of. Although neither dealer nor owner will lie if asked directly about the animal's disposition or soundness, neither will he volunteer information detrimental to the sale. Cheating at horse trading has never borne the stigma of dishonesty that other types of cheating bear!

Occasionally one has "a friend" or a "friend of a friend" who has "just the animal" for sale. For some reason it is impossible for him to keep his horse so he wishes to find a good home for him and will

accept a very reasonable figure. Beware of such bargains, many is
the lifelong friendship which has failed to weather such an adven-
ture, for the animal that may be "just the thing" for him may not be
at all the thing for you. At least don't commit yourself until you
have had a chance to try out the animal thoroughly and have him
vetted.

Of all places for the amateur to go for horseflesh, the AUCTION SALE
is the worst. About the most that can be said for the horses that are
put up for auction is that there is something the matter with them
or they wouldn't be there. Somebody had no further use for them
and, being unable to sell them in any other way, has sent them to the
auction sales with the hopes of realising a little something on them.
I do not mean to imply that no good horses ever go to auctions, but
certainly it is not the place for any except the knowing. Devious are
the means of making a horse step out sound and show some life for
the few minutes that you have to make up your mind about him,
and it is you who will take the loss if he later turns out to be unsound
or otherwise disappointing. To be sure many sound, young, "green"
horses are shipped in from the west each year and go for very small
sums at the sales, but there is no way of trying out such a horse and
you will be able to tell little about his disposition, manners or gaits.
Better let the dealer buy him and then examine him at leisure at his
stable even though you may have to pay a bit more.

There is yet another place where horses are to be bought, and that
is at the YEARLING SALES at the race tracks. Here the colts which are
not fast enough for racing are auctioned off. Beautiful animals often
go for small sums, but it takes the expert to distinguish between the
good, sound colt and the one with weaknesses not so obvious to the
amateur.

HOW MUCH SHOULD ONE PAY?

It is impossible of course to give a definite answer to this question:
how much should one pay for a serviceable animal? So much depends on the age, breeding, conformation, etc. of the animal in question as well as on the time of year, locality and market. But you
should be able to find a useful, well broken "family" horse for from
a hundred and fifty to three hundred and fifty dollars. A hunter or a
jumper will cost more, especially a heavy weight. If you live in the

west or south you may get a local, farm or ranch-bred animal for a great deal less. In horses, as in everything else, there are no real bargains. You get just about what you pay for, but some slight blemish that would keep a high class animal out of the show ring may very well bring his value down to fit the moderate purse. Lack of condition which can be remedied with care and feed, may also make it possible for the person who does not want to pay the fancy prices to get a really good animal.

TESTS FOR SOUNDNESS AND MANNERS.

In buying your horse try and get permission to have him on trial for a week, failing this, go several times to ride and drive him. Try him out both in company and alone. Make sure that he is willing to leave the stable without a fight, there is nothing more annoying than the "barn rat." Watch him being saddled and bridled, he may be headshy. If he is to be driven, watch him harnessed, you don't want to buy a beast that needs one person to hold his head and another to run the shafts up on him. Park him parallel and also "heading in" to the curb and see if he backs out without getting excited.

Drive or ride him in traffic and past unusual objects. Remember that the modern horse is more afraid of the pheasant that flies up under his feet, or the child on roller skates, than he is of a steam roller. An animal that will stand without tieing, or one that will stand tied to a portable weight is much more useful than the one that needs some one to hold him, or which depends upon a bona fide hitching post.

If possible take a veterinary with you. Failing that check the eyesight by seeing that the eyes are clear with no discharge and without a bluish tinge, that the horse blinks when you wave your hand *behind* the eye, and that he does not carry his head to one side.

To check soundness of legs, have him ridden towards you and away from you at a trot on a hard road. Notice if he travels straight or if he tends to point in or out with his toes. The horse with the former habit is usually a stumbler and the one who toes out may have to be ridden in boots to keep him from knocking his ankle or cannon bone. Listen to the beat of his feet, the cadence must be even, not accented. The latter indicates that the horse is favouring one foot. To test shoulder lameness ride him down hill at a trot

Pick up his foot and see that the frog is well developed, the sole cupped, not flat.

Run your hand slowly down each leg for its entire length both to feel for blemishes and to test the animal's manners. Have the owner bring him out of the stable cold, saddle and warm up in front of you. Certain types of lameness show up only before the blood is circulating fast. Horses with bad wind will often show a hacking cough at the beginning of the ride and will get over it after a few minutes, another reason for viewing the animal before he has a chance to work out of his troubles. To further test his wind, gallop the horse and then look at his flanks to see if they are heaving. Do not confuse the noise that some horses make by blowing through their nostrils with the laboured breathing of the windbroken animal.

Choose the horse with a short barrel, well-spread ribs, plenty of girth and rounded hind quarters. The rangy, leggy, loosely put together animal will cost you three times as much to feed and always look half starved. A narrow chested beast has not room enough inside him for a good heart and lungs.

In selecting a pony take along a child to ride him. The pony that cannot be managed by a child is of no earthly good no matter how appealing he may be. Manners are all important in the child's mount, test him thoroughly for kicking and biting. Take plenty of time in choosing the right animal, be it pony, hunter, combination or hack. Remember that a young, sound horse will be good for ten or fifteen or even more years of work. Select him wisely but don't expect perfection, you won't get it.

CHAPTER II

Selection and Care of Equipment

STABLE EQUIPMENT.

In addition to a saddle and bridle you will need the following for your stable; a saddle pad, unless your saddle fits perfectly, a halter and two or three lead ropes (halter shanks), a stable blanket, a summer sheet and a wool rug for day use in winter, these last two are optional. *Grooming tools* are as follows: a dandy-brush made of coarse fibre to get the heavy dirt off. Soak this brush in a pail of water overnight before you use it the first time. A body brush (hair) a curry comb, rubber if you can get it, the round metal type is preferable to the kind with teeth if the rubber is unavailable, a rubbing rag, a mane-and-tail comb and a hoof pick. A body-scraper is optional. The best are of flexible metal with leather handles, one can be improvised from a barrel stave.

The dandy-brush costs anywhere from sixty cents to a dollar and a quarter. The body-brush will be more, perhaps as much as two-fifty. The mane-and-tail comb can be bought for fifty or sixty cents. You may be able to improvise a hoof pick out of a heavy screw hook, by simply screwing the latter into a small block of wood, or the blacksmith can turn you out one. The rubber curry-combs, which sell for about sixty cents are much better than the metal for two reasons, first, they won't break or bend if stepped on, but mainly because you may use them on your horse with impunity as they cannot scratch him. The circular metal type can be used on the body of the horse with care, the type with rows of teeth is good only for cleaning the brushes.

In the stable you will need several large, heavy, galvanised watering pails. Put one of these aside for First Aid work and keep in it your medical supplies for treating wounds (see chapter on First Aid), then, if you want your things in a hurry they will all be together. You will often use Lysol for washing wounds, and as a horse

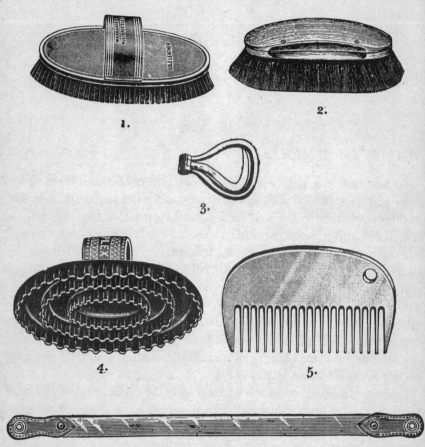

Grooming Tools

1. Body Brush, 2. Dandy Brush, 3. Hoof Pick, 4. Rubber or Metal Curry-Comb (spoon type), 5. Mane and Tail Comb, 6. Body Scraper.

will not drink out of a pail that smells of Lysol it is necessary that one particular pail be kept for that purpose.

For cleaning the stable you will need a rake, a broad-bladed, square-edged shovel, a fibre push-broom a manure fork (four or five tined) and a wheel-barrow. In the feed room you will want a two or three quart measure (a sauce-pan with a handle does nicely and is far cheaper than the standard wooden measures) and a hay fork. If you get your hay in bales keep a pair of wire snippers hung close by.

The well dressed stable-man needs blue-jeans and work shirts. A leather jacket, smooth finish, is preferable to a cloth one for cold weather. It doesn't retain the smells, is waterproof and hairs do not stick. Loose work gloves will save blisters and callouses. Gardening gloves, treated with a hand softening cream, are particularly good. Boots to use in the stable are a problem. Riding boots are not satisfactory. They are hard to get on and off, uncomfortable for walking and it is difficult to pull blue-jeans over them. Although heavy, ski-boots are better, especially in wet or snowy weather. Rubber boots, if big enough to take felt linings or several pairs of woollen socks inside are good as they can be hosed off and don't absorb odors. In summer any comfortable, old shoes or sneakers kept for this purpose will do.

Keep a hammer, wire snippers, wrecking tool and jack knife where you can get them in a hurry if a horse ties himself up in wire or gets cast in a stall. The knife with the leather hole-punch in it serves two purposes.

Electric clippers cost around twenty dollars but they last a long time, are a great convenience and the blades can be resharpened or replaced as worn. Stables and hunt clubs charge from five to seven dollars to clip a horse so by doing your own you very soon save the cost of the machine.

For care of tack you will want saddle-soap and some Neats-foot oil or Neats-foot compound. Black harness will need either black harness oil or black harness soap as well as metal polish. Get the kind that comes in cans at the hardware store. There are many varieties of saddle soap. Most common is the kind that comes in a round can like wax shoe polish. Certain automobile soaps made of vegetable oils are also excellent for leather. This soap is of a jelly like consistency and works into the leather well.

For applying the soap you need several sponges known as "tack sponges" and available at harness stores. They are much tougher than the very cheap sponges to be found elsewhere. Cellulose sponges are inexpensive and very satisfactory for this purpose. A big "sheeps wool" sponge, though expensive is good to have around for sponging off your horse in hot weather. For mending harness a little gadget known as a "speedy-stitcher" is useful for emergency repairs although it will not do as neat a job as will the harness maker.

TYPES OF SADDLES.

There are two general types of riding saddles, the "ENGLISH" or flat saddles and the "WESTERN" saddles. Included in the English type are the so-called "PARK" or "HACKING" saddles, the FORWARD-SEAT or jumping saddle and the RACING saddle. The latter is very small and light and useful only on the track. The park saddles are the kind most usually seen. The forward-seat saddle was introduced a few

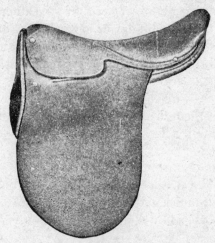

Hacking Saddle

years ago, it differs from the park type in that the skirts slant forward and that the cantle is slightly higher which makes it easier for the rider to keep his forward position in the saddle. This type is also built with knee rolls to assist the rider in keeping his knees in place in jumping. If you can get a forward seat saddle by all means do so, but they are harder to find and more expensive. All of the types of saddles mentioned above are for use in riding horses with a regulation trot (when the rider wants to "post"), for polo, for racing, for jumping and for showing in any except "Branded Horse" classes. The principle of this type saddle is to put as little extra weight on the horse as possible and at the same time make the rider comfortable in any position which he may need or care to take. It would be of

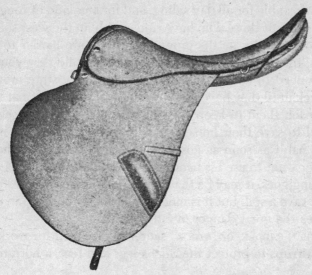

Forward Seat Saddle

no use in roping cattle, nor would the cowboy's stock-saddle be good for jumping, racing or polo. One hears the argument that the Eng-

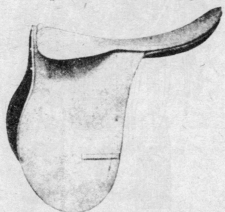

Racing Saddle

lish saddle is not suitable for riding in the west because the rider is expected to ride all day and would get too tired in the flat saddle, also that the rough and steep trails make the other type imperative. Any one who has done any hunting knows that a flat saddle is perfectly comfortable for all day riding, and for any kind of rough country. Beginners, or those who have ridden only in the west sometimes ask for stock saddles on the ground that they are harder to fall out of or that "the horn is so nice to hold on to." While very young children may need to hold on to the pommel or to a strap around the horse's neck until they have gotten a modicum of balance at the trot, both to teach them to keep their hands down and themselves forward, and to save their horse's mouth, it is well to get them away from this habit as soon as possible. No rider, worthy of the name, would ever want to use his hands on his saddle to retain his seat! As far as falling goes, it may be that the deeper seat of the western saddle might save a spill but it is much easier to get hung up in one and therein lies the great danger in falls.

THE WESTERN SADDLE, or "STOCK" saddle is designed for work. It has covered stirrups to protect the rider's legs and feet, a horn to which

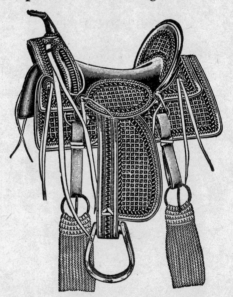

Western Saddle

is attached the lariat, and a high cantle to which the blanket roll is fastened. It is built high off the horse's back and a very heavy, folded blanket is used to protect the spine of the animal as much as possible. It is fastened by means of one or two girths (cinches) which are tightened by pulling on leather straps and so can be made tighter than would be possible with the buckle type of girth; this is necessary because of the tremendous pull put on the saddle when the horse is used for roping. It is easy to get these saddles so tight that they injure the horse or prevent his breathing properly. Western ponies often develop a habit of swelling up and hardening their muscles when girthed, as a result of having been cinched up too tight.

THE OFFICER'S FIELD SADDLE is a cross between the western and the flat saddle, being a flat saddle mounted on a "split tree" with a fan tail that sticks out behind and to which the blanket roll can be

Officer's Field Saddle

attached. It is a very useful type of saddle, especially for a big horse with extra high withers, but it is heavier than the regulation flat saddle.

THE MCCLELLAN SADDLE or Army saddle is a western type with only one cinch girth and a modified horn. It is light, cheap, easy on the horse as it has a split tree but is not suitable for show riding, polo, jumping or ordinary hacking.

It is important to learn the names of the various parts of the saddle and bridle (see diagram, page 22) if only to be able to specify what you want in buying.

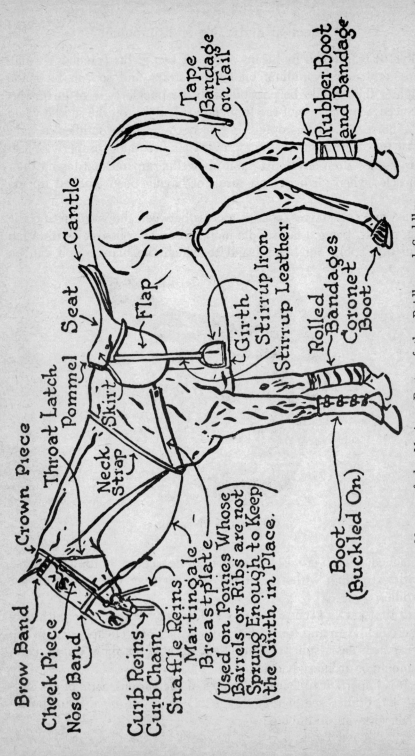

Brow Band
Crown Piece
Cheek Piece
Throat Latch
Nose Band
Pommel Seat — Cantle
Skirt
Flap
Neck Strap
Curb Reins
Curb Chain
Snaffle Reins
Martingale
Breastplate
(Used on Ponies Whose
Barrels or Ribs are not
Sprung Enough to Keep
the Girth in Place.)
Girth
Stirrup Iron
Stirrup Leather
Boot
(Buckled On)
Rolled Bandages
Coronet Boot
Tape Bandage on Tail
Rubber Boot and Bandage

Names of the Various Parts of the Bridle and Saddle

BUYING THE SADDLE.

New saddles of every type and price may be found anywhere from mail order houses to the more expensive sporting goods house, but most horsemen prefer to buy a used saddle because a saddle needs to be broken in just as does a pair of shoes. Horse dealers often have such saddles for sale as do hunt clubs, riding stables and the bigger harness shops. Wherever you go for your saddle and no matter what you pay, bear the following points in mind:

For a horse with prominent withers a "cut back" throat and a well stuffed saddle is essential. If possible try the saddle on your horse before buying; if you can see from front to rear between the horse's spine and the saddle you won't need a saddle pad. Try to get a saddle that is leather-lined, wool linings wear out in a few years, they are hard to keep clean and are subject to moth holes.

GIRTHS.

The most satisfactory type of girth is that of folded leather. The "split" girth is cooler and for the horse that gets girth sores there is a special girth made of braided leather (see illustration, page 24). The webbing girth is quite satisfactory but it will not last as long and is a good deal harder to clean.

STIRRUPS.

Most stirrups are nickel plated to prevent rusting. They may have a roughened tread to help prevent slipping of the boot, or they may have rubber pads, the latter are nice in cold weather as they make for warmer feet. Get your stirrups amply wide so that there is no danger of getting your boot jammed and stuck. The stirrup leathers should be flexible and strong, the better ones are numbered to make it more convenient for adjusting the stirrups evenly. The iron in the flap of the saddle where the leather attaches should be of the "safety" type, then it will pull out easily in case of accident and eliminate the possibility of your being dragged. Be sure to sit in the saddle yourself before buying. It is as important to get one that fits *you* as it is to get one that fits the horse. All harness and sporting goods stores have wooden "horses" for this purpose. If you buy a

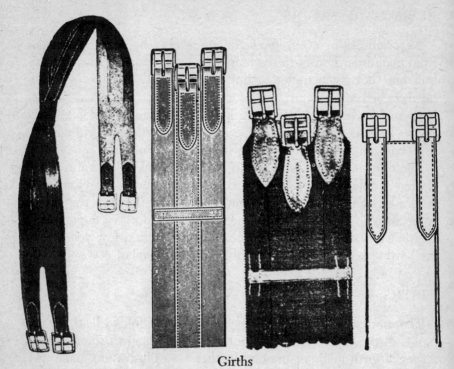

Girths

Leather and Web Six Buckle, Fitzwilliam Girths, Folded Leather Girth

new saddle, oil it well before using. It will take away that "new" look that is the brand of the tenderfoot and will also prevent it from darkening unevenly. Neats-foot oil is good for this purpose, some people use kerosene.

Second-hand saddles run from forty dollars up, a really good new one will cost from a hundred to a hundred and fifty.

THE RIDING BRIDLE.

The type of bridle you buy will depend largely on your horse and for what you are going to use him. Find out from the previous owner what he is used to and make your purchase accordingly. If you can afford a new bridle do so, as bridles wear out in a few years whereas a good saddle, well taken care of, will last a life time. However a good, second-hand bridle will last quite a while if you treat it with respect.

The test of good leather is its flexibility and a feeling of "life." Don't buy leather that looks the least bit cracked or dried out, it isn't safe. There are three types of fastenings for bridle parts; hooks, buckles and stitching. The hooks are supposed to be neater but they are much harder to manipulate. The buckles are cheaper and more convenient. The stitched bridles are the most expensive, if you get one of those you had better have a spare handy, for if anything breaks it means a trip to the harness makers before you can use your horse again.

If you have small hands don't get reins that are too wide. The braided reins used on hunting bridles are splendid for wet weather, but this type is never used on any except the snaffle bridle.

Learn to take your bridle apart and put it together again without confusion. The general tendency is to get the bit in backwards so that the reins hang to the front instead of to the rear, or to get the cheek straps twisted. Once you get a clear picture in your mind of what the bridle looks like you will have no more difficulty.

FAMILIAR TYPES OF BRIDLES.

SNAFFLE BRIDLE. The simplest bridle is the SNAFFLE BRIDLE (see illustration, page 26). This is used on horses with light mouths. It is used on most hunters and is best for beginners so they do not have to worry about learning to handle the double reins. It is also wise to use this type on jumpers, especially colts, as the damage to the horse's bars (the spaces between the back and front teeth in a horse's mouth on which the bit rests) is considerably less if the rider comes back on the reins over the jump, than with a Pelham or full bridle. The snaffle bridle is not satisfactory for showing in saddle or horsemanship classes, nor for use on a horse which holds his head too high (though sometimes a snaffle with a running martingale can be used on such an animal), nor on the beast with the very hard mouth.

As is shown in the illustration on page 26 the snaffle bit is jointed and has a ring at each end to which the reins are fastened. The bit should be adjusted so that it just wrinkles the corners of the horse's mouth and works on the lips rather than on the bars as do the Pelham and curb bits. With a confirmed puller its action is simply to raise the animal's head rather than to stop him which explains why it is not satisfactory for this type of horse.

Familiar Types of Bridles and Bits

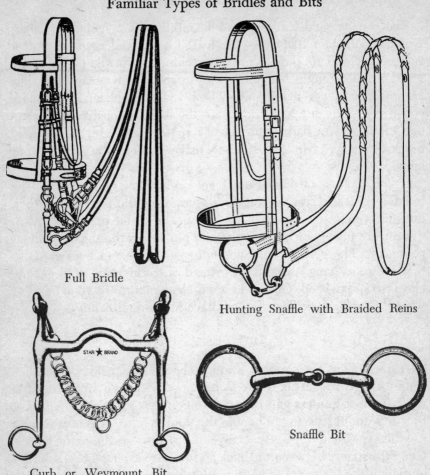

Full Bridle

Hunting Snaffle with Braided Reins

Curb or Weymount Bit
with Port and Chain

Snaffle Bit

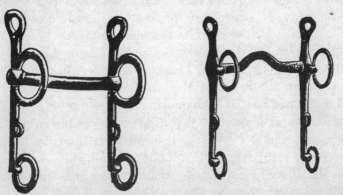

Pelham Bits

THE PELHAM BRIDLE is used frequently in polo and for hacking. As you see by examining the illustration on page 26 it is sort of a combination snaffle and curb. Many authorities maintain that it is entirely wrong in principle unless the rider has such good hands that he never pulls both reins at the same time, as by doing so he raises the bit into the position of a snaffle, yet the action is that of a curb, the chain pressing much too high on the jaw. This is true in principle, yet many horses work better in Pelhams than in anything else. THE FULL BRIDLE is a snaffle and a curb put together. When used in this way the horseman speaks of the two bits as being a bit (curb) and a bridoon (snaffle). The latter is placed as it should be, fitting snugly into the corners of the horse's mouth, while the curb rests an inch or two lower on the bars. The action of the curb is quite different from that of the snaffle. When the curb reins are pulled the bit is pressed down on the bars, the curb chain is pressed against the chin groove so that the horse's lower jaw, at this point, is pinched between the two. The effect of this is to lower the horse's head and cause him to bend at the crest instead of raising his head as does the snaffle. A severe jab with the curb is real cruelty, and some horses are so afraid of this type of bit that only a very good horseman can manage them in one.

Most experienced horsemen prefer the full bridle to other types, and they are a necessity in showing in horsemanship or good hands classes.

Learn to adjust your bridle so that it is comfortable and to notice instantly, if something is wrong, a twisted curb chain or a too short cheek piece can mean agony to your horse.

New bridles cost from seven to twenty dollars, depending on the quality of the leather, they are usually sold without bits which must be purchased separately. Second-hand bridles can be had for considerably less. Darken the new bridle as suggested for the saddle. THE CAVESSON was originally designed to prevent the horse from opening his mouth too wide when the reins are pulled and so getting away from the discipline of the bit. It consists of a narrow strap going around the nose which is held in place by another strap which goes over the head and behind the ears. This strap runs through the loops of the brow band as does the crown-piece of the bridle and is placed under the latter. It should be adjusted so that the nose-band comes two inches below the horse's cheek bones and so that you can slide two fingers under it but no more. Some bridles come equipped with a simple nose-band attaching directly to the cheek straps.

THE MARTINGALE.

There are two types of martingales, each with its specific purpose. By studying the illustration below you will notice that the so-called "STANDING MARTINGALE" is simply a straight piece of leather with a loop at one end through which the girth passes, and a smaller loop at the other which is attached to the cavasson or nose-band.

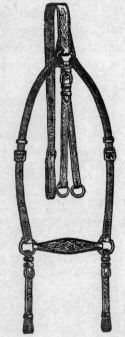

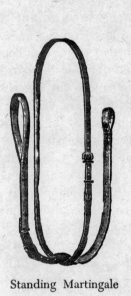

Standing Martingale

Running Martingale with
Hunting Breastplate

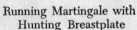

Martingales

The purpose of the standing martingale is to steady the horse's head and particularly to cure him of the trick of tossing his head up suddenly and catching the unwary rider on the forehead. It may be used in jumping if not put on too short, but adjusted so that it allows the horse to extend his neck freely. It is held in place by a breast-plate encircling the neck just in front of the shoulders. Hunting breast-plates have little straps attaching to the saddles, others are simple lengths of leather which should not fit too snugly. The RUNNING MARTINGALE fastens to the girth in the same manner as the standing type and it too is supported by a breast-plate but the other end is split with two short ends each ending in a ring (see page 28). The snaffle reins are then run through these rings. The purpose of the running martingale is to give the rider more control with the snaffle. It is adjusted so that when the horse holds his head in a normal position there is no pull on the reins, but if he throws his head up and tries to avoid the bit, instead of the latter working on the corners of his mouth, it is now pulled down and works against the bars, tending to depress the head and chin. Many horses that are incontrollable in a snaffle without a martingale go splendidly in one. Be sure that it is short enough to be of some practical use. This type of martingale is excellent for use on an animal that tends to hold his head too high while jumping; needless to say it should never be attached to the curb reins.

THE STABLE BLANKET.

Your horse will need a stout, warm stable blanket day and night in cold weather, or if you can afford it, he can be really well dressed and wear the stable blanket at night and a warm wool rug in the day time. The former costs from three or four dollars to ten or twelve. It should be of sturdy duck or canvas (burlap is not satisfactory, it tears too easily) and be lined with some warm fabric. It is sometimes possible to buy used army blankets for as little as a dollar and a half, these are strong and serviceable although they do not stay in place as well as other types, especially those made with one broad surcingle in the center of the back and a slide type of buckle with a web strap across the chest. If the blanket seems to be too loose at the shoulders and hangs too low in front, take a deep tuck over the withers.

Blankets come in several sizes, sixty-four to seventy-four inches in length, measuring from back to front at the hem. The blanket should come back almost to the horse's tail and the hem should hang at the line of the belly, if too long the animal will step on it and tear it whenever he gets up. Two surcingles are preferable to one, they should be adjusted so that they fit snugly but not as tightly as you would fit a girth. One or two straps in front with snaps or buckles keep the blanket together across the chest and some blankets come equipped with crupper-straps to go under the tail as well. The wool rug is usually a colorful plaid, they are expensive, but certainly improve the appearance of your stable and they can double as a lap robe in the sleigh.

THE STABLE SHEET.

The stable sheet is put on the horse in the daytime in warm weather. It protects him from flies and dust and keeps his coat down. A new one costs ten or twelve dollars, but you can easily make one for two or three dollars and a couple of hours' labor. Use any durable fabric, the porous type such as crash is cooler, although denim is very satisfactory. You will need about three yards of twenty-eight inch material and seven or eight yards of wide tape binding. Use the same tape as tie straps across the chest, a surcingle may not be necessary, if it is, one can be made from the tape. Your stable blanket provides the pattern.

HARNESS.

If you own a combination horse you will need harness. Either russet or black is suitable although the former is associated with country driving and the latter with more formal occasions and vehicles. New harness is best, but expensive. A serviceable plain single harness can be bought for about fifty dollars. Pony harnesses are less, thirty-five to forty. Second-hand harness in good condition is hard to find. Be sure that the reins, traces and hold-back straps are perfectly strong. These must not break in an emergency. Your final selection will be limited by the size and weight of your horse and by the type of vehicle he is going to draw as well as by what you are able to find.

If you are not familiar with harness, study the diagram on pag 32 until you know each piece and its purpose. To a novice a har ness is just a collection of complicated straps, to a horseman each strap has its particular use and one glance tells him exactly where it belongs. Before buying consider the following points:

First the bridle, most horses go well in what is known as a "BLIND

BRIDLE," i.e., one with "blinders" or flaps which prevent the horse from seeing behind him. A few do not like such a bridle. The blinders should not flop nor should they hug the horse's eyes so closely that ninety percent of his vision is cut off. It is more difficult to adjust the cheek straps for size in a driving bridle than in a riding bridle, hence more care must be taken in selection. The bit is adjusted just as is the riding bit, see page 25. Some horses go well in snaffles, others are more easily handled in what is known as a "Liverpool" bit, see page 32. This bit allows for three adjustments as to reins, the lower you attach the rein, the more severe the bit. A horse that has never gone in a curb will not work well in a Liverpool. The highly schooled horse, accustomed to a full bridle and trained to take a canter when collected on the curb will probably go better in a snaffle also. Be careful in adjusting the chain of the Liverpool to see that it is flat and that you can get three fingers under it without disturbing the position of the bit.

THE HARNESS PAD AND BREECHING are fastened together by the BACK STRAP which ends in a CRUPPER that fits under the tail and prevents the harness from sliding too far forward. The pad has terrets, metal rings through which the reins pass. It rests directly behind the horse's withers and is held in place by a girth. The breeching (pronounced "britching") is what the horse sits back against when going down hill. THE HOLD BACK STRAPS run from the breeching to the shafts of the vehicle. Usually these are rewrapped each time the horse is harnessed, and left attached to the breeching when the harness is not in use, but occasionally one sees them permanently fastened to the wagon with snaps to hook onto rings in the breeching. Although this saves time in harnessing it is not a safe practice, the straps may rot where they bend over the shaft and their condition not be noticed. Also readjustment as to length is hard and it is very easy to forget to oil them inasmuch as they are separate from the rest of the harness.

Certain types of carts are made with special irons on the shafts to prevent the vehicle from riding forward on the hills and so make breeching and hold-backs unnecessary. Green horses often find the breeching hard to get used to, resting as it does on the hind quarters not far above the hocks. If your horse is in a straight stall you can get him accustomed to the feel of the breeching by keeping a chain or strap behind him in the daytime.

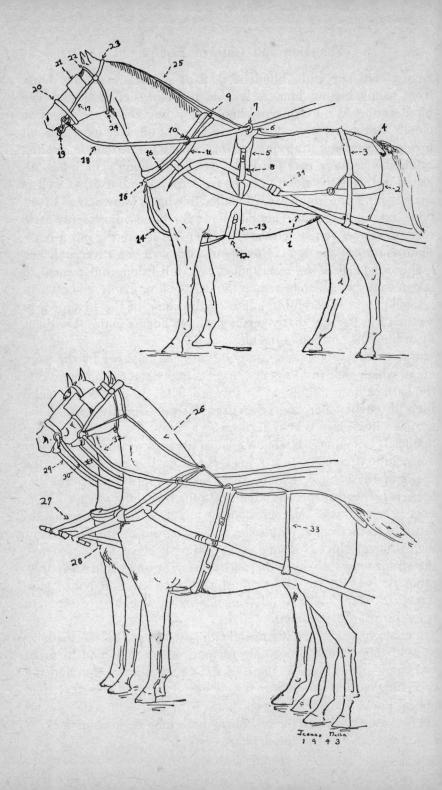

Jeanne Mellin
1 9 4 3

Single and Double Harness (*see illustration on page 32*)

1. Trace, 2. Breeching, 3. Kicking strap, 4. Crupper, 5. Backband, 6. Pad, 7. Terret, 8. Tug, 9. Hames strap, 10. Hames Terret, 11. Collar, 12. Belly-band, 13. Girth, 14. Breastplate, 15. Kidney link, 16. Hames, 17. Cheekpiece, 18. Reins, 19. Liverpool bit, 20. Noseband, 21. Blinkers, 22. Browband, 23. Crown piece, 24. Throat lash, 25. Overhead check, 26. Side check, 27. Pole strap, 28. Pole, 29. Off rein, 30. Off coupling rein, 31. Near coupling rein, 32. Near rein, 33. Loin strap, 34. Hold back. Notice that in double harness no hold back strap is worn as the horses hold back by means of the pole straps attached to collar and pole piece.

COLLAR.

There are two types of collars used in driving. One is the KAY collar which is a big, shaped collar of leather with a hard frame and upon which rest shaped metal rods known as HAMES. To the hames are fastened the TRACES and the other end of the traces fasten to the single-tree on the wagon. By pushing against the collar, which protects the horse's neck from the hames, the horse draws the vehicle. The collar *must* fit the horse. Otherwise you will have a very uncomfortable animal with badly galled shoulders and withers. It should be padded in such a fashion that it does not press on the top of the withers, and long enough so that it does not touch the windpipe at the bottom of the neck. The width of four fingers is the usual measure of the space between the windpipe and the bottom of the collar. The lining of the collar, which is also a source of galled shoulders, must be kept scrupulously clean. The hames are fastened together at the top with a strap and at the bottom with either a link known as a kicking link or a kidney link, or by a chain. They must fit the collar and must be put on tight enough so that there is no chance of their pulling off. A BREAST-PLATE goes around the collar and hames and then between the horse's legs ending in an adjustable loop through which the girth passes. Its purpose is to prevent the collar from sliding forward on the neck.

Another type of collar is the DUTCH COLLAR, sometimes called the "breast collar," see page 35. This is a simpler arrangement, being merely a broad strap with traces attached which is held in place by a narrower strap going over the horse's shoulders. It does away with the necessity of the hames and consequently of a heavy, padded collar. It is suitable only for very light work.

The TUG, through which the shaft runs, is fastened to the pad by the BACK-BAND (don't confuse with "back-strap"). The position of the tug and back-band will tell you whether or not the traces are too long or too short; if they swing in front of or behind the side of the pad your horse will be pulling his load on the back-band instead of on the traces.

The mention of a CHECK-REIN makes the average person think immediately of "Black Beauty." True, the check-rein, wrongly adjusted, can be an instrument of torture. Equally true, a properly adjusted check-rein will cause no discomfort and has prevented many an accident as it keeps the horse from putting his nose down to his toes and kicking or bolting. A properly adjusted check-rein (bearing-rein is a better term), allows the horse to hold his head naturally with no pull on the mouth unless he drops it too low. There are two types, the over-head check and the side check. The latter is not as severe as the former and may be used successfully on well trained horses.

VEHICLES.

Roughly speaking, vehicles may be classified as follows, two-wheeled vehicles (carts and gigs), and four-wheeled vehicles (wagons, runabouts, surreys, phaetons, victorias, etc). Inasmuch as there have been no new vehicles for a number of years, carriage makers having turned their shops into factories for automobile bodies, you will have only used vehicles from which to choose. Pick out something sturdy, you can always dress it up with a coat of paint. Vehicles that have been allowed to stand under cover will often have loose spokes due to the wood drying out, this can be remedied by soaking the wheels thoroughly several times. Dealers who trade in horses will sometimes be able to supply wagons and carts as well. The sporting section of the Sunday editions of the newspapers carry advertisements of either privately owned vehicles or establishments where an assortment is to be found.

Many people prefer the two-wheeled type of vehicle for the following reasons: they are strongly made, they do not tip over as readily and are much easier to turn at sharp corners or in a narrow roadway. On the other hand they have their disadvantages. They are heavier to pull both because of the heavier construction and because the shafts are in one piece with the body, putting more weight on the tugs, and they have a rough, jogging motion.

TYPES OF CARTS.

Every one is familiar with the GOVERNESS CART with its door at the rear and two seats facing each other. The body may be either of wood or of wicker, when made of the latter it is known as a "BASKET CART." The old fashioned "TUB CART" is a country version of the same thing. These carts are low, sturdy and very safe, but they do take a strong pony as they are heavy when loaded with an adult and several children. They bring from a hundred and twenty-five to two hundred dollars.

The "Tub Cart"

This is a home made version of the more stylish "governess" cart. The pony is a small hackney, not too highly bred to be useful.

The DOG CART is a high, two-wheeled vehicle, designed originally to carry the hounds to the hunt. Space for the latter was provided under the seat in a box-like compartment, setting the driver well above the horse. The COCKING CART, whose purpose was to transport fighting cocks is built on the same principles but even higher and more dangerous. If you plan to drive a sporting high-stepper or a tandem you will need one of these carts, otherwise they are not practical for the average horse or driver. Incidentally many people do not know that the practice of tandem driving was originated by the hunting men of England, those wonderful sports who thought nothing of riding a dozen miles to the meet and hunting all day, faced with the prospect of an indeterminate distance of hacking home, depending on where the last fox was killed. Wishing to spare their horse, they devised the means of driving him ahead of a driving horse. The latter did all the work of pulling cart and master while the hunter trotted ahead between slack traces. To make it even more practical he built his cart to accommodate his hounds.

Thoroughbred Filly Hitched to a Breaking Cart
Notice that no breeching is necessary.

The MINEOLA cart is lower than the dog cart with seats that fold back. It is a nice, useful type for country work. There are various types of BREAKING CARTS. The true breaking cart has very long shafts with special irons making breeching unnecessary, and an extension running out behind so that if the horse rears he cannot go over. There are many modifications of this, that, if you can find them are cheap, safe, and serviceable though not stylish. Some sort of breaking cart is essential if you are breaking a green horse or trying to turn your saddler or hunter into a combination.

The SULKY, used on the trotting track, is light with bicycle wheels. It is lots of fun if you own a speedy trotter but not very practical as it holds only one passenger and no packages. Carts come with or without rubber tires, the latter make for a quieter ride but are hard to replace when worn.

FOUR-WHEELED VEHICLES.

The most common type of four-wheeled vehicle is the RUNABOUT. This comes in a variety of styles. It may be all of wood or may have some wicker in the body. It may or may not be "cut-under." In a

Hackney Ponies Hitched to a Runabout

These are not show ponies and so are not too high strung to make excellent combination animals for a child of ten or twelve years.

cut-under the body is cut away so that when the buggy is turned the front wheels go under instead of hitting the guards on the side, hence a shorter turn may be made and there is less danger of tipping over. If you can get a runabout with a fringed parasol you will be both elegant and comfortable.

The TOP BUGGY, most familiar of all in the country in the days when the horse ruled the road, is higher and not so stylish as the runabout, but it is very practical for bad weather driving. The top folds back for good weather and with it up and a rubber apron-sheet, easily improvised from an old poncho or a length of oilcloth, in front of you, you can be snug and dry in bad weather. This type tips over very easily and is recommended for use only with a well-broken, gentle horse. The DOCTOR'S PHAETON shown below is a rich relative of the country top-buggy.

Western Chunk Hitched to an Old-Fashioned Doctor's Phaeton

This type vehicle and animal are ideal for the family who want something quiet and comfortable.

If you have a large family you may prefer to buy a carriage that will hold four people. There are a great variety of these, from the country BUCKBOARD with the "jump-up" seat to the family SURREY with the fringed top. You will need a somewhat stronger, heavier horse for this type of vehicle, a "western chunk" would be splendid or perhaps a pair of combination horses or ponies. A vehicle designed to carry four should have a brake, especially if you live in hilly country, it saves your horse and also your nerves. Upholstery on wagons and carts may be of leather (strong and serviceable but how hot it gets if left out in the sun!), or of any fabric which will stand the weather. Whipcord is good, broadcloth is often used on the more formal vehicles.

To go with your vehicle you will need a lap-robe, heavy in winter, light for summer. You may use this to throw over your horse to keep him warm or protect him from flies or you may carry his own rug or sheet along. If your horse is the type that will stand tied to a portable weight you will need one of these, they weigh from twenty to forty pounds and are attached to the bit by means of a halter shank and clip. Adjust the rope so that the horse cannot pick it up off the ground when he raises his head, and so that he cannot get his feet tangled in it. You will want a buggy whip for your family type vehicle, a more sporting type whip for your dog-cart. Perhaps you may never need to use a whip, but when you want it you want it badly and it is never safe to go without.

As far as price goes, there is no standard scale. Breaking carts and top-buggies are the cheapest, more formal vehicles and two seated affairs start at a hundred and fifty or two hundred.

CARE OF HARNESS AND TACK.

Daily care: Wipe all leather with a damp sponge and saddle soap. If it is muddy or covered with dried sweat, scrub it first, using a brush or a wet sponge and plenty of soapy lather. Rinse under running water and then with an almost dry sponge work the soap into the leather without lather. When it becomes necessary to moisten the sponge this is best done by dipping your hand into the water and dampening the sponge with it rather than by dipping the sponge itself directly into the water. Be sure to pay particular attention to billet straps, stirrup leathers, girths and the hard, high part of the cantle.

Bits, stirrups and curb-chains may be brightened with scouring powder, steel wool or sand and water. An old fashioned method is to pull up a clump of grass and use the roots with the soil that clings to them. Metal parts of the harness such as the terrets should be kept shining with polish.

Once a month undo all buckles on the bridle and harness, take out the bits, take the girth and stirrup leathers off the saddle and soak everything made of leather except the saddle itself, the harness pad and the Kay collar in a tub of Neats-foot oil or Neats-foot compound. Allow to remain in the oil until the leather has absorbed all that it will, then remove it and wipe dry with a clean cloth. Wipe the heavy parts that are too thick to soak thoroughly with plenty of oil, working it well into the cracks between the linings and uppers of collar and saddle.

Old leather that is mildewed or dried out may be somewhat restored by washing with soap and water and then oiling as suggested above. For daily use black harness soap is less messy than the harness oil usually sold, although many people like the latter.

EMERGENCY REPAIRS.

The point at which the bridle is most apt to break is where the cheek-strap fastens to the crownpiece. A horse, tied by the reins, or allowed to step on dragging reins, will invariably break his bridle at this point. If only one side is broken, let the other side out one hole and fasten the broken side in the hole above the break. If both sides are broken let out the cheek-straps a half inch or so where they fasten to the bit; meanwhile order a new crownpiece. Sometimes the only damage is to the tongue of the buckle which must be straightened.

Broken reins can be mended on the road by putting one end over the other so that they both lie flat, boring two holes an inch apart, through the double thickness of leather and threading a shoe-string or a bit of wire through the holes. Later they can be stitched at this point. This same method of repair can be used on any of the lighter straps of the harness, it is much safer than trying to tie the broken ends.

If a trace breaks where it fastens to the single-tree you may be able to let it out at the collar and put in new holes.

Most saddles have three billet straps, if two break you can still get home by attaching one buckle of your girth to the third. Only a harness shop can repair billets as the saddle must be taken completely apart.

Sometimes the stitching which holds the various buckles, particularly those on the stirrup leathers, will part. Cut an elongated hole through the leather where it was bent to go over the buckle. Now run the tongue of the buckle through this hole and hold the flap down in place until the buckle is fastened. It will be just as strong as before and you can have the stitching replaced at your leisure.

If you are not going to use your harness or tack for some months, oil it well and store it in a dry place; leather is very subject to mildew and dampness.

CHAPTER III

The Stable

LOCATION.

A stable need not be either unsightly or smelly. It must be built of good materials or you will have to rebuild in a year or so. It must be designed conveniently or daily chores will take more than their share of your time. It must be located near enough to your house so that you can hear if something goes wrong during the night, far enough away so that you will not be bothered with flies and it should be attractive to look at.

The stable should be on well drained ground and with a sunny exposure. If there are two rows of stalls facing in opposite directions, the east-west exposure is good, if all stalls face one way the southern exposure is best. The adjoining paddock should be to the east or south. If you are building a new stable you may want to tie the architecture in with that of your house.

MATERIALS.

An excellent material for outside walls is the asbestos shingle that comes impregnated with white so that it never needs to be re-painted. It is attractive, water-proof, fire-proof and a good insulator. It comes in several different sizes and styles and is no more ex-pensive than wood shingles, clapboards or siding plus paint. Inside the stable all stall partitions and wooden flooring *must* be of two inch oak boards. Nothing else will take the hard wear and banging. Grills between stalls may be of iron or they may be of heavy, gal-vanised, diagonal mesh wire.

SPECIFICATIONS OF STALLS.

BOX STALLS. *The standard size for box stalls* (often called "loose boxes"), is twelve by twelve. A medium sized horse gets along nicely in a ten by twelve or even a ten by ten. Locate your feed manger and your water pail or fountain in opposite corners as the latter become easily clogged with scattered grain and then emit a very bad odor. Some horses like to "dunk" their hay, regardless of where the water is placed, the bowls of such animals will have to be washed out frequently with a strong solution of soda.

STRAIGHT STALLS. These take up less room than the boxes but are not as comfortable for the average horse as he will have less freedom of movement. However a great many horses do very well in straight stalls and some few horses do better than when housed in a box stall where they become restless and hard to handle.

Straight stalls should be of a size to accommodate your horse. A horse of fifteen hands or thereabouts needs a stall five feet wide and nine or ten long. A big hunter should have one at least six by ten while a pony does nicely in a four by eight. It is usual to have the planks of the partitions removable rather than fixed so that if a horse should splinter one it can be replaced without taking out the whole section. Special irons come for this purpose or you can achieve the same effect with wood. A practical method is to use the irons for the front of the partitions and then build posts at the other end. Incidentally, every inch of wood that has an edge and can be reached by the horse's teeth should have that edge covered with sheet metal or it will be chewed to bits. Cover the front five feet of the top board of each partition before you put on the grill, the box stall partitions will have to be covered their full length. Planking on partitions should go up a distance of five feet with a grill above to permit circulation of air.

FLOORING OF STALLS.

The best flooring for a box stall is firmly packed clay with a layer of crushed rock beneath for drainage. This also goes for straight stalls if the slope is right, the drainage adequate and the clay the type that packs hard. Clay is best for the horse's feet and it is a deodorant. Sand is bad as it may work in behind the walls of the horse's feet and cause trouble. Wood soaks up the moisture and retains the smell but it is next best to clay. Cement is very bad. It is cold and damp as well as hard. A horse that stands too long on cement is prone to develop tendon and foot troubles. Some horses delight in undertaking private excavations at night which means that one must fill up the hole again in the morning, tamping it down firmly but this is really the only disadvantage to clay.

If you are remodelling an already existing building for a stable, one which has a cement floor, about the best thing to do is to build a hinged, wooded rack which can be raised for cleaning purposes. This will not suck up the moisture as much as the solid board floor and hence will not retain the odors.

STABLE ACCESSORIES.

FEED MANGERS should be of iron. If you must use wood, line the box with sheet metal and see that all edges are thoroughly protected. Use IRON RACKS for hay in straight stalls. They should be hung at about the level of the horse's head, if hung high the dust from the hay will often cause inflammation of the eyes and lead to more serious eye conditions. Hay is best fed on the floor but this is not practical except in loose boxes where the horse is free to retrieve the hay that he scatters. In a straight stall much will be pushed back under the hind feet and wasted. The old-fashioned type of straight stall where the hay manger is built in across the front of the stall, the feed box being in one corner is good, but takes up a great deal of space.

Individual WATER FOUNTAINS are best if you can afford the additional piping. Use the automatic type designed for use in a cow barn, horses learn to work them readily and they save an enormous amount of labor and time. They also prevent the spread of infectious diseases. Their only draw-back is that in a cold climate they do occasionally freeze at the top where the paddle joins the head, but a little boiling water from the tea-kettle thaws them out promptly. All exposed pipes must be insulated except in warm climates. If you want individual water and do not want the expense of the automatic fountains, you can nail a board across the corner of each stall and slide a pail into it. This means more work than the former arrangement but it is less trouble than leading each horse out of his stall for water from a common trough.

Another arrangement, where stalls face each other, is to build a sloping trough between with a faucet at one end and a drainage outlet at the other. This is a simple and inexpensive way to solve the watering problem but it does not have the advantage of keeping down the spread of contagious diseases such as shipping fever.

In addition to the water bowls or other arrangements for watering your horses you will need a sink where you can wash tack and where you can draw water in pails or attach a hose. By locating this sink near a window with an outside trough in the paddock, below said window, it is possible to use the same outlet for outside watering as well. As it is sometimes necessary to hose a horse either because he has become over heated or to reduce inflammation, have this in mind when you locate your outlet.

TIE ROPES AND CHAINS.

If your horse is to be kept in a straight stall he will probably have to be tied at the head to prevent his trying to turn around in his stall. It is possible simply to attach one end of a rope to his halter and the other to a metal ring set about three feet from the ground. In this method there is always the danger of a horse putting his front leg over the rope and then throwing his head up and getting caught. An excitable horse can hurt himself very badly in this fashion. A better method is illustrated in the diagram on page 70. A ring is set in the wall, a small block of wood is then attached to a chain or rope which has a snap at the other end. Run the end with the snap through the ring from the floor and attach to the halter. The weight of the block will keep the chain from getting slack and at the same time the horse can get the full length of the chain when he wants to lie down or reach upwards. It is convenient to have rings with chains and snaps on either side of the straight stall at the rear and on either side of the door of the box. You can then tack up your horse ahead of time, cross-tie him in his stall and he will be unable to get the reins under his feet or to roll in the saddle. You will want to put similar tieing arrangements where ever you plan to do your grooming. Light, strong steel chains do not kink and last longer than the more common hemp ropes but the latter are perfectly adequate.

In addition to the chain or rope at the horse's head you may find it advisable to keep something behind him if he is in a straight stall. This, to some extent, prevents his trying to kick at another horse or at a person passing behind him, and curbs the tendency of backing out of the stall suddenly (see vices, page 106). About the best thing to use for this purpose is a length of chain or rope run through a bit of old rubber hose. If the horse comes back against this suddenly it will not injure him. It should be so located that it hangs a few inches above the hock level. Many horses will stand in straight stalls with a chain of this description and nothing else to hold them in. A few will not countenance a rope behind them as they get their tails over it and go frantic.

LATCHES AND HANDLES.

Latches and handles especially on box stalls, would be of the type that cannot be opened by a horse, as well as very strong. Many horses will learn to slide an ordinary bolt so be careful what you buy. There is a type with a rounded handle, upon which the horse cannot get a grip with his teeth, that is very efficient.

Another useful gadget is one that catches and holds the top half of the box-stall door open. It consists of a spring, set on the wall with a pointed metal arrangement on the door which slips into the spines and is held.

LIGHTING FIXTURES.

Lighting fixtures should be placed high and out of the way so that there is no danger of a horse hitting his head. Be sure and have a plug outlet for use with the electric clippers, or else install one socket which holds both bulb and a plug. Switches should be near the door and located where they cannot be injured by the horse. Don't forget plenty of light for the hay-loft.

VENTILATION.

Stables should be light and well ventilated but a horse should never have to stand with the light in his eyes, nor should he have a draft blowing directly on him. This can be avoided by having the windows placed high and hinged at the bottom so that they open outwards from the top. In a stable where there is a row of box stalls with half doors the tops of which may be kept open in hot weather it is a good plan to have such windows on the opposite wall. Sliding doors at either end then permit a circulation of air from all directions; in cold weather, by closing all doors, direct draughts are avoided.

THE TACK ROOM.

This should be near the stalls. It should have running water and a sink with enough bridle and saddle racks to take care of all your tack. A portable saddle rack on which to put your saddle while cleaning it is a great convenience. Bridle racks should be set high enough so that the reins can hang straight down without having to be doubled. You will need an iron hook attached by a short chain to the ceiling on which to hang your bridle while you soap it. If you have a driving harness you will need separate racks of special types for that. All these racks are expensive if bought new but they can sometimes be picked up second hand or they may be improvised from wood or cans. If there is no space in your stable for a separate tack room, hang your tack to the left of your horse's stall. It will be harder to keep clean because of the dust from the hay, but this method has the advantage of being convenient particularly where there are a number of horses.

THE HAY LOFT.

Hay and grain are most conveniently kept in a loft directly over the stalls. Cut a square opening over each hay manger so that the hay can be forked down. Put in a three inch iron pipe leading through the floor of the loft down into the feed box. By using a large funnel you can now shoot the grain directly into the box without going into the stall at all. This is a tremendous time saver and is also a safety factor as some horses kick if you go into their stalls while they are eating.

It is absolutely essential that you have a rat-proof container for your grain. If you have only one horse or pony and buy your grain a bag at a time, a good sized metal garbage can with a tight lid, is as practical as anything else. You will need two, one for oats and the other for bran unless you use an already prepared horse feed, when only one will be necessary. If you get your grain in quantity you should build a wooden box with a hinged top and cover it with fine, square-mesh wire. Cover it on the *outside* or the rats will gnaw through the wood, the grain will seep through the wire and your labor will be wasted.

Your loft should have a ventilator set in the roof as without one there is great danger of overheating and consequently of fire.

PADDOCKS AND PASTURES.

A small paddock, adjacent to the stable is essential. It should not be muddy and twenty or thirty feet square is adequate. There should be drawbars between this and the pasture so that if your horse is difficult to catch or if you want to turn him out for only a short time while you do up his stall, you will have him where you can lay your hands on him. A pile of sand or straw in one corner gives him a good spot on which to roll and scratch his back without getting dirty. In the Italian army the horses are taken to a sand-pit each day and allowed to roll as part of the routine. A water-trough should be available. It is sometimes possible to pick up a tub or sink suitable for this purpose, at a second hand store that handles such things or at a plumbing shop. Wash-tubs are excellent.

The FENCING surrounding the paddock and pasture may be of sheep hurdling, panel fencing, wire (not barbed wire, unless you want your horses badly marred) or a stone wall. If the latter you may want a bar on top of the wall. Drawbars or a gate for your horse, a turnstile or a narrow opening between the post and the stable wall for yourself will save time. Speaking of drawbars, some animals, especially ponies, are past masters at taking these down. They will pick them up in their teeth and slide them, or get their withers underneath and raise them and the next thing you know you get a call from an irate neighbor. The only really satisfactory answer is to attach a short chain with a snap to the bar and a heavy staple to the post and keep this fastened at all times. Wire fences will last longer and not be such a potential danger to your horse if a strip of ribbon wire is fastened to angle irons which in turn attach to the posts. These should be set in such a fashion that they hold the band of wire a foot in from the main fence and a few inches higher. Electric fencing is said to be very satisfactory. It may be necessary to train your horse to respect it. This is done by fencing off a small corner of the paddock with two strands of the charged wire. Now place a bucket of oats on the far side. In reaching for it the horse will receive a shock and soon learn to keep away. These fences are very cheap to install but must be inspected frequently to make sure that no weeds or twigs come in direct contact with the wire and so ground it.

A pasture with good grass saves hay and benefits your horse. It should have some shade. If it is not very large or you have several animals, divide it into two by a central fence and turn your horses into only one section at a time, giving the other a chance to grow. You can also utilise your pasture for riding by having a track with or without jumps around the edge, leaving the good grass in the center. Keep the horses in the small paddock in the spring until the grass in the pasture has gotten a good start, and don't turn them out in muddy weather or the turf will become cut up. Liming the soil once a year in the fall or early spring will kill off the weeds and bring up the clover.

CARE OF MANURE.

The most satisfactory way of handling the manure problem, provided you have a horse or pony that will pull, is to have a little dump cart and keep it backed up outside the stalls. Then as you clean out said stalls, wheel the wheelbarrow directly into the cart and empty it. When the cart is full which may be daily if you have a number of animals or only every few days if you have only one, haul it well away from the house and dump it in a pile. If you have no use for it yourself it will bring a good price provided it is kept in a neat heap and not allowed to become scattered. If you have no animal that is broken to harness, locate your manure-pile a convenient distance from the stable and arrange with some nursery, farmer or commercial manure company to collect the manure weekly. You will thus largely avoid the fly nuisance.

CONVERSION OF A GARAGE.

There are two main problems in converting the average garage into a suitable stable for a horse or pony. The first is the flooring which is usually cement. This can be overcome by building racks as suggested on page 44. If for any reason this is not practical the horse must be given plenty of bedding, straw to his knees, if straw is used; six or eight inches of shavings if you prefer the latter and the same for peat moss or peanut shells. The second problem is that of ventilation. Most garages are built with sliding or overhead doors containing fixed windows which give light but no air. These will have to be replaced with some type of window which can be raised or opened. The type suggested on page 46 is hard to install because of interference with the working of the door. It may prove cheaper and more satisfactory to build entirely new windows on the side walls.

Few garages have storage space above them for hay and feed. If it is a two car garage that is being converted this may be overcome by partitioning off a section to be used for this purpose and for grooming, handling and storage of tack, etc. Be especially careful of your choice of latch which closes your box stall door or to the chain which holds your horse in a straight stall, many an animal has met an early death by getting into the oat bag. It will save you steps if you arrange to have your feeding manger on the partition between the stall and the space allotted to grain and hay with an opening in the latter you can thus feed and hay your horse without getting into the stall.

CHAPTER IV

General Care

PRINCIPLES OF FEEDING.

The stomach of a horse is small in proportion to his body, for this reason a horse should be fed comparatively small amounts of grain at a time but fed often. In his natural state a horse lives on grass and he is eating all the time that he is not resting. He is not required to do any strenuous or hard work; horses in pasture seldom run for any distance except to escape enemies, so the grass is sufficient to keep them in fair condition although they will not be hard and if put to work will sweat easily. But the point is that under natural conditions a horse eats slowly, digesting as he eats. Grain, however is another matter, it is a concentrated food, hard and with a tendency to swell when wet. The stabled horse, fed only two or three times a day, tends to eat his grain too rapidly, it all gets into his stomach at once and stays there while being digested—a process that takes from one to two hours. The teeth of a horse are flat and intended for grinding, the horse chews with a circular motion, his digestive juices are not strong and nature intended him to thoroughly masticate his food before swallowing it, furthermore a horse has no vomiting muscles and so cannot get rid of anything that once gets into his stomach and disagrees with him. This is one of the main reasons why so many horses die of acute indigestion and fodder poisoning.

Water, on the other hand, taken on an empty stomach, goes right through into the kidneys, therefore it is easy to see that the time to water a horse is *before* and not directly *after* he has finished off a good feed of grain, for, going into the stomach after the grain, the water will cause the grain to swell and the horse may have an acute case of colic as a result. Water given on top of a feed may also work undigested oats into the intestines. Hay is eaten more slowly and it is perfectly permissible for the horse to be allowed to drink his fill after hay. It is best to put the hay in the rack *before* you put the oats in the manger. The horse will take a few bites and be less apt to bolt his oats. To summarize, feed and water your horses in the following order; water, hay, oats.

AMOUNT AND KINDS OF FEED.

The amount and kinds of feed necessary to keep a horse in good condition depends on several factors, on his weight, on his general type and on the kind and amount of work he will be expected to do. The light driving or saddle horse averages a thousand pounds in weight. The heavyweight hunter twelve or thirteen hundred, the draught horse from eighteen to twenty-four hundred. In addition to this it must be remembered that, like people, some horses apparently assimilate their food much better than others and so stay fat on a far lighter diet. In horses this seems to depend partly on the breed and also on the build. A short-barreled, chunky type of horse keeps fat on very little, while the long, rakish animal looks lean no matter how you feed him. Racing horses, whether thoroughbreds or standard breds are usually hard to keep fat, while western ponies are notoriously "easy keepers." Horses that are to be used for fast work such as polo ponies, racing horses, hunters, etc. will need more oats and not as much hay whereas horses that do slow work such as draught horses will need less oats in proportion to their size but more hay. The United States Cavalry ration provides from nine to twelve quarts of grain daily with from twelve to seventeen pounds of first quality hay. Cavalry horses are expected to do from four to six hours of work, averaging a total of twenty-five miles, carrying a man and pack and going at the rate of six miles an hour. A horse trots nine miles an hour, gallops twelve miles and walks four, the cavalry breaks up its marches into fifty minutes of traveling, ten minutes rest and varies the gaits which rests both horse and man. In addition the horses may be called upon for forced marches of as long as a hundred miles in one day. Cavalry horses are usually half or more Thoroughbred the remainder being saddle,

Morgan, or western strains. These horses are given from nine to twelve quarts of grain a day, depending on type and work, and from twelve to twenty pounds of hay.

Three feeds a day are better than two unless your horse is doing very light work or is turned out during the day. Colts, particularly future race horses, are often fed four times a day. Ponies, especially Shetlands, do better with very little or no grain, and the western cowboy lets his mount live more or less on the land; but he changes mounts frequently and most of his work is slow, "movies" to the contrary. Divide your feed as follows: morning; one third the ration of grain, a little less than half the hay—noon; a handful of hay if desired, another third of grain—night; remainder of hay and grain.

GRAINS.

Oats—Oats are muscle builders. They give a horse pep. They should be thin skinned, hard, sweet and weigh forty to forty-five pounds a bushel. They may be fed whole or crushed. If crushed they are more easily digested and assimilated but care must be taken in purchasing crushed oats as it is harder to judge of their quality. Some dealers take the oats that are not saleable whole, and crush them, selling them as first grade oats. Crushed oats have greater bulk per pound than whole oats and so more should be given if you measure by the quart. Many stables keep an oat-crusher and crush their own oats.

Bran—Bran is a laxative. It has a little nutritive value but not much; however most horses do better on a mixture of bran and oats than on oats alone. The usual proportion is three parts oats to one bran in summer, two parts oats to one bran in winter measured by bulk. Bran is heatening and horses fed too much bran in hot weather may break out with eczema. Some prefer to feed the bran once a week in the form of a mash. Take four quarts of bran, add a handful of of salt and enough boiling water to wet thoroughly, cover with a lid and allow to steam until cool. Feed before a day of rest. Linseed meal may also be added.

Corn—Corn may be fed either on or off the cob. It is fattening and very heating, especially to horses not accustomed to it. It is used in the west and south more than in the east. The army has been experimenting with a combination of corn, oats and bran, one third of each, but they feed it only in cold weather or when the horses are not doing fast work. Fed on the ear the usual ration of corn is five to nine large ears. Thus fed it is good for the gums and teeth, horses fed on corn do not need to have their teeth floated as often as others and some authorities believe that it helps to prevent "lampers" a condition in which the gums and roof of the horse's mouth swell. Horses that have never been fed corn will often refuse to eat it entirely, others eat not only the corn but the husk as well!

Linseed Meal—This is not really a food but if given in small amounts it has a good effect on the bowels and makes the coat glisten. A handful three times a week mixed with the other grain is the right amount.

Commercial Horse Feed—Sometimes called MOLASSES FEED. This is a prepared feed being a combination of oats, corn, molasses, chopped alfalfa, linseed, salt and other minerals. It is fattening and many stables prefer it, but it is too heating in hot weather and better for horses doing slow work than fast as it is not such a good muscle builder as oats or oats and bran.

FRESH FORAGE.

Carrots—Very good for the coat, stimulates the horse's appetite. Break up into small enough pieces so that the horse will not choke on them.

Apples—Loved by horses but not as good as carrots; many a case of colic has developed from Dobbin getting loose in the orchard.

Grass—The horse's natural food. A horse that has not had fresh grass for some time, such as one coming out of the stable in the spring, should be allowed to get used to grass gradually. Horses troubled with eczema will usually get rid of it as soon as the grass comes in. Grass is supposed to contain more vitamins than any other form of vegetation but it also contains a great deal of water and is not as much of a muscle builder as oats, nor as good a fattener as corn. If you want your horse to have grass but cannot turn him out, you may give him grass that has been cut, but feed it at once, don't let it lie on the ground. Grass that is cut and allowed to lie without being properly cured will ferment in the horse's stomach and may cause colic or fodder poisoning. Be careful never to use grass or allow a horse to graze where there is any danger of contamination from the spraying of trees. This spray is deadly poison. For this reason, if for no other, it is unwise to allow a horse to snatch a bite of grass when you are riding. Too much grass causes gas and gives a horse what is known as a "grass," or "hay belly"; his sides stick out, yet his ribs show and there is little meat on his rump.

Hay—Forms the bulk of a horse's diet. It should be short, fine, smell sweet, and crackle when you crush it between your fingers. It should be greenish in color rather than straw colored. If the latter it means that the hay has been cut too late and has very little nutriment. Horses will lose more weight on bad hay than from any other cause. It is poor economy to buy cheap hay, not only will you have to feed more grain, but it will not go as far and your horses will lose condition no matter how hard you try to keep them up. Most hay used for horses is a mixture of timothy, red top and a little clover. Many people claim that alfalfa makes a good feed for horses. It is far richer than the other types of hay, more laxative and a little mixed with the regular hay promotes growth, but horses that are to be fed straight alfalfa must be accustomed to it gradually, or they will scour badly and develop skin troubles. Never feed hay, grass or grain that has any form of mould on it or that has been allowed to get wet and stay wet. Horses that tend to respiratory troubles or show signs of coughing when fed should have the hay wetted down just before it is fed to them.

SALT.

All horses need salt, especially in hot weather as in sweating the horse loses a great deal of the body salt which must be replaced otherwise the horse will suffer from heat exhaustion. Lack of sufficient salt also accounts for the loss of weight sometimes noticed in the summer months. Salt for horses may be bought either in bricks, in which case you will need a special dispenser for it, or it may be bought in big lumps looking like grey rocks, in the latter case just put a piece the size of your fist in the manger along with the grain, it will also tend to keep the horse from bolting his grain.

COST OF FEEDING.

It is hard to make any absolutely accurate statement as to how much it costs to keep a horse but I have found that twenty dollars a month will cover his actual food and his shoeing. This is in the East where all food must be bought and where the horse is expected to do a fair amount of work every day.

WATERING.

A horse will drink from five to eighteen gallons of water a day depending on the weather and how hard he is working. Horses that are not given sufficient water lose condition quickly and may develop kidney troubles. There are two recognized systems of watering, both good. One is to have water before the horse at all times. This may be done either by having the automatic water bowls mentioned on page 44 or by putting a pail of water where the horse can reach it but cannot tip it over. Description of how to install such a pail is given on page 44. It may seem paradoxical to say that a horse should not be watered directly after being fed, and then to say that he should have water before him at all times, but the answer is that a horse that has water before him whenever he is in his stall will never get very thirsty, consequently he will only drink a few mouthfuls at a time. Therefore it is safe for him to have access to both his water and grain at the same time. If the horse has been out of the stable for some time, allow him to take his drink when you bring him in and then give him his grain. There is only one word of caution, a hot horse should never be allowed to drink his fill and then stand. This may bring on the condition known as "founder" (see page 81). But no hot horse should ever be allowed to stand, water or no water. He should be cooled off gradually, rubbed down, and blanketed if the weather is cold. It will do no harm if a hot horse is allowed to drink *provided* he is kept moving afterwards. Never keep a fagged-out, thirsty horse from water, it is cruelty, let him have a drink and then walk him in slow circles until his circulation has quieted down and he is cool to the touch. The place to test a horse to see whether he is hot is between the front legs on the brisket. The cavalry, when on the march, makes a practice of camping a mile or so *beyond* the water hole. The troops water their mounts and then dismount and walk in to camp so that the horses have a good chance to cool out and need only to be rubbed down and put away. There are two advantages to keeping water before the horse in individual pails or bowls, the first is that you are sure your horse will not go thirsty, and the second is that it cuts down the spread of contagious diseases such as shipping fever.

The second satisfactory method of dealing with the watering question is to water four times a day. One to two hours after the morning meal (few horses will drink before breakfast if they are at all hungry), before the noon feeding, before the night feeding and the last thing before you go to bed. In watering a horse in this way from a pail, a stream or a trough it must be remembered that horses like to drink slowly unless very thirsty indeed. This dates back to prehistoric times when the horse, like the deer, had to keep a weather eye open for enemies that were apt to lie in wait at drinking pools for the unwary. So even today your horse will drink a few sips, raise his head and look around and then drink a few more. Don't hurry him, let him take his time. When he takes his head away from the water and starts searching for a few wisps of grass or a spilled grain you will know that he is through. It goes without saying that a horse cannot drink comfortably in a bit although he may appear to do so. If you are on your way home from a ride and want your horse to drink his fill, slip the bit out of his mouth but keep the reins just behind the ears holding them in your right hand close to the throat. Have your left hand ready to grab his nose if he tries to duck away when he is through.

BEDDING.

Your horse should have a comfortable, dry bed at night. Some horses lie down, some do not. Tests have been taken recently which show that a horse rests better on his feet, be that as it may a horse should certainly be able to lie down if he wants to and his bed should not be dirty, wet or cold. Furthermore if his bedding is scanty he may injure himself in trying to get to his feet again. If enough litter is not provided to absorb the moisture you will have a hard time keeping him clean.

There are several kinds of bedding available, each has its good and bad points. Straw is the most common. Rye straw is the best, oat straw is unsuitable as the horse will eat it. Some horses will eat any kind of straw and will have to be bedded on something else. Straw that is chopped up will be more absorbent than any other but few stables have a chopper. Each morning you must shake out the straw, dry clean straw may be left in the stall, dirty straw and manure is taken out and put on the manure pile, if wet it may be put in the sun to dry out and put back in the stall at night together with additional straw to take the place of that which was removed. Stalls with cement floors need much more straw than stalls with wood floors, and the latter need more than stalls with clay floors. Box stalls should have the bedding distributed evenly over it, straight stalls may have the bulk of the bedding at the rear with only a little up under the front of the horse. SHAVINGS make a very satisfactory bedding. They absorb moisture readily, tend to reduce the smell and no horse will eat shavings at least not more than a mouthful or so. Furthermore if you live near a lumber-mill you may very easily be able to get your shavings for nothing. Four grain bags filled with shavings is ample to take care of a box stall for a week, three will do for a straight. Clean the stall thoroughly, getting out every bit of litter, and put in two bags of shavings, spreading them around the stall, this covers it to a depth of three or four inches. Each morning after that take out any manure or wet bedding, and add just a little new litter, at the end of the week again clean stall thoroughly and start over. With the straight stalls, rake out all litter each morning, leave the stall bare during the day and add fresh shavings at night. A sugar scoop is very handy for handling the shavings if you only want to use a little at a time. The only disadvantage to shavings is that the manure is not as valuable for use as a fertilizer as the shavings do not decompose as quickly as other forms of bedding, so you may not get as good a price for the manure, but if you can get your shavings for nothing this will more than make up the difference in price. Fine shavings are best and if you can get pine or cedar shavings you are in luck. Use old oat bags to hold shavings. They are easily handled this way.

PEAT MOSS is favored by some stables. It increases the value of the manure, is very absorbent and keeps down the smell. Many people do not like it because it mixes with the manure making it difficult to tell one from the other and it gets in the horse's coat; also, it is expensive.

PEANUT SHELLS for use as bedding are available in certain localities. They have proved satisfactory for use on poultry and dairy farms, horses might be inclined to eat them. They cost about as much as shavings.

SHREDDED SUGAR CANE is also recommended as bedding. It decomposes more readily than the shavings, is deodorant and absorbent but rather expensive for a large stable. Above all, don't use hay for bedding. Farmers will sometimes try to sell hay that is no good for fodder on the ground that it will do for bedding. The horse will eat it, dirty or clean, and will get digestive or respiratory troubles and worms as a result—besides using up the bedding before he lies down. On a pinch, hay may be used under the hind-quarters of horses in straight stalls if it is not put where they can reach it with their mouths.

LEAVES. Dry leaves, raked up in the fall, make fairly satisfactory bedding although they are so light and crush so easily that it takes a good many. Furthermore, because of this bulk factor, storing is difficult. Whatever bedding you use, just be sure that it is clean, dry and sufficient in quantity.

GROOMING.

"A good grooming is worth a feed of oats," is also a saying long familiar to horsemen, and the truth of it is obvious because the grooming, acting as a massage, stimulates the circulation by bringing the blood to the surface. Nourishment, being carried to the tissues by means of the red corpuscles, the horse assimilates and uses the food he eats to a much greater degree if he is thoroughly groomed every day. The muscular exercise of grooming is equally good for the groomer, for, properly done, grooming is a strenuous exercise.

There is a third benefit to be derived from grooming. By going over every inch of your horse daily, you will quickly notice anything wrong with him and be able to check it before it becomes serious. Of course there is also the more obvious reason for not neglecting your grooming as in so doing you greatly detract from your horse's appearance.

Having tied your horse where there is plenty of light and where you can move around him freely, and having placed all your grooming tools within reach, pick up your rubber curry comb in your right hand and your dandy brush in your left. Stand on the near side of the horse, opposite the neck. Brush his face and head, not forgetting the poll. Use only the brush and use it gently, some horses are fussy about having their heads groomed, others like it. Next rub his neck back and forth, round and round with the curry comb. This brings the dandruff and dead hair to the surface. Follow by a vigorous brushing with the dandy brush, working with the hair. Use a sweeping motion and lift the brush at the end of each stroke. Don't dab at your horse, put your back into it and really work. When you have finished with the neck go on to the belly, barrel and brisket. Then the back and rump. On the legs below the knee use only the brush. Be sure you get the pasterns at the back below the fetlock joint thoroughly clean, mud is apt to stick here causing irritation and may develop into a condition known as "scratches." When you have finished one side go to the other. If your horse has been neglected or been "on the rough" it will take you a good many groomings to get him really clean. Clean your brush after every few strokes by combing it with the curry-comb, clean the latter by tapping on your heel or against the stall. When working on the right side it is easier to hold the curry-comb in the left hand, the brush in the right. When you have finished with the dandy brush take the body-brush which is of soft hair and polish him with this. Next take up the mane and tail comb. A horse has no nerves at the roots of his hair so don't be afraid of hurting him. It is easier to untangle a mane or tail if you start at the end and comb a few inches at a time. If you find any burrs, disengage them by pulling the hairs away from the burrs, not the burrs away from the hair. Horses in pasture in the fall get little yellow specks which look like seeds on their legs. These are the eggs of the bot-fly and should be cut off. Otherwise they may get into the horse's intestines where they will hatch into grubs.

If you have trouble making your horse's mane lie over one side you can train it by wetting it thoroughly, dividing the offending section of hair into bunches, taking about three inches to a bunch, and braiding it into tight little pigtails on whichever side you want it to lie. Leave it this way for a week or so without undoing it.

Many horses wear their manes hogged (clipped short). This simplifies the grooming problem, but one never clips a Thoroughbred's mane nor a five gaited saddler's and it is usual to clip only the forelock of a trotter or harness horse. The saddler should be allowed to grow his mane as long as it will (this includes the Tennessee Walking horse as well). For shows the forelock may be divided into three straight braids which are tied with bright colored wool the ends of which hang down. The Thoroughbred and the trotter have their manes shortened and thinned by pulling until it is about four inches long. This is a tedious process. If you have a pair of tooth edged thinning sheers such as the beauty parlor uses, you can give your horse a fair approximation of the proper hair cut, or, if you are handy with a razor, wet the mane, divide into strands and feather cut it from underneath. A horse with prominent withers and a scrawny neck looks better with a mane, a horse with a coarse, heavy, short neck looks better without. It is a good plan to leave the manes on ponies or horses which are to be used by beginners, it gives them something to hold on to when they are practicing riding without saddle, jogging or having a try at jumping.

In regard to tails, the saddler's is left long and sweeping, the Thoroughbred wears his thinned and shortened to just below the hocks and the polo pony and cavalry horse often have theirs banged (cut straight across). Don't ever use the clippers on any tail except a mule's. In thinning or shortening, pull the hairs out in little bunches of two or three hairs, it won't bother the horse.

After you have finished with the mane and tail go to the feet. Take up the hoof-pick with your right hand and, starting with the near fore-feet, lift each foot in turn and clean the sole and frog thoroughly. This will also give you a chance to inspect the shoes and see if your horse is in need of new ones or if his toes are getting too long. Be sure to face the rear whether you are working on the front or the back foot. Facing the front while working on the front foot invites disaster as many a horse will take that opportunity to plant his back foot in the middle of *your* back and send you sprawling. In cleaning the back feet, to which some horses object strenuously, you will get a better hold if you bring the hoof between your knees as the blacksmith does. If your horse is reluctant to pick up his feet lean against the shoulder with your shoulder thus throwing the horse's weight on his opposite foot, now run your hand down his tendon to the fetlock and pull up on the hair, speaking to the horse as you do so. It is easy to train a horse to readily lift his feet simply by insisting that he do so and not giving up when he resists. Do not clean out the feet at night, only in the morning, the earth and clay that your horse picks up when he is out will help to keep the frog and sole soft and moist, this

When picking up a forefoot, the rider faces to the rear with her shoulder against the shoulder of the pony to shift his weight to the opposite foot. Notice that the stirrup is run up on the leather so that it will not dangle and possibly hit the rider if the animal makes a sudden motion.

is especially important if your horse is standing on wood or cement even though he may have plenty of bedding. However, it is just as well to look at the feet at night to see that there are no stones and pebbles in them.

Lastly, take up your body rag and wipe off any dust on your horse, finishing with a good hand rubbing. Your horse should now be immaculate! If he isn't, it means that he has not been properly groomed for some time. If you want to test whether or not he is really clean, rub your fingers through his coat against the hair, grey tracks indicate that he still has dandruff or dust in his coat.

SHOEING.

Good blacksmiths are becoming more and more rare, for it is a craft that is hard to learn and takes diligence, strength, experience and above all pride in a job well done. Lazy, careless blacksmiths as well as ignorant ones ruin many a horse. In the old days the craft was passed down from father to son, certain families often making a specialty of some particular type of horse, for horses are shod according to the work to which they are to be put, thus a saddler or a fine harness horse wears heavy shoes and longer toes to increase the showiness of his action, race horses wear light, racing plates, just heavy enough to last the race while trotters and pacers have to be shod exactly so to make them extend themselves to the utmost and get the greatest distance out of each stride without breaking into a gallop. Horses used on snow or ice wear shoes with special pegs to keep them from slipping. The horse that overreaches, grabbing his front heels with his back toes must have his front toes shortened so that he can pick his front feet up quicker and get them out of the way, while his shoes behind should extend a little to slow up the action of the back feet. Horses that brush their ankles with the opposite foot have to have shoes specially weighted and shaped to cure this as do horses that toe in or out. A horse whose shoe pinches will develop corns just as does a human, or a condition known as "contracted heels" which causes lameness. By the above you will see that shoeing a horse is not just tacking a piece of iron or shoe on the bottom of his foot, and one wonders where the new blacksmiths are to come from when the present generation that grew up and were trained before the day of the automobile are gone. Fortunately the army trains a number of smiths and the race tracks and trotting tracks are an incentive to some others to keep going.

Though few horsemen will be able to do their own shoeing everyone should know when a shoe fits and when it does not and the basic principles of good shoeing. First, why does a horse need shoes? In his natural state he does not wear shoes and his feet are in good shape. True enough, but in his natural state a horse is not called upon to draw nor carry weight, he does not travel fast enough except in an emergency and he does not work on hard ground. The outside wall of the foot is made of the same material as your fingernails, like your fingernails it keeps on growing so that when broken or worn off it replaces itself. Obviously the horse running barefoot will keep his own feet worn down at least to some extent, but the horse whose feet are protected by shoes must have the shoes removed at least once a month and the extra growth cut off. Unless this is done the horse will stumble and, furthermore, the wall having grown long, the frog no longer comes in contact with the ground. The purpose of the frog, that raised triangular shaped object on the bottom of the foot, is to act as a cushion and help absorb the shock when the horse travels. Properly shod the frog still touches the ground, improperly shod or with walls too long all the shock must be absorbed by the rim of the wall. It goes without saying that each shoe must be fitted to the shape of the individual horse, not, as so often happens with careless or lazy blacksmiths, the foot rasped down to fit the shoe. Below is a diagram showing a hoof that has been well shod and also one in which the outer wall has been rasped down incorrectly.

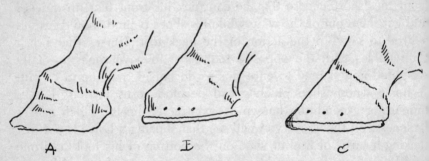

A   This foot indicates a condition of "Founder" or Laminitis. Notice the ridges on the wall and the concave appearance.
B   A hoof that has been badly shod, the wall having been rasped away to fit the shoe.
C   This is a well shod hoof.

Before fitting the shoe the blacksmith will trim the foot and level it so that the horse will walk squarely. It is here that the greatest skill is required and it is in this leveling and trimming that many a horse's foot is ruined when the smith arrives. Mention any trouble you are having with your horse's action. Ask him to look for signs of thrush. If the sole tends to be hard rather than chalky you might have him pack the foot with tar and oakum. If your horse is tender in front due to an old injury, or if he has a slight tendency to laminitis, he may go better if you have him shod with leather or rubber pads. If you do much driving on hard roads you will certainly need the pads.

In winter you may want shoes with detachable pegs called "sharp shoes." Keep the pegs in only if your horse is travelling constantly on snow or ice. Under no circumstances should he be allowed to work on dry roads with sharp shoes except at a walk for the shock then comes on the points of the shoe only and will soon cause a strain to the tendons or other injury.

The cost of shoeing varies greatly with the locality. Three-fifty to five dollars for machine made shoes is average in the North and East. More for hand made ones. In the South and West the prices are much less.

CLIPPING.

It is hard for many people to understand why a horse is clipped in the fall, just before cold weather sets in. The answer is very simple. The horse grows a heavy undercoat in the winter. If he is worked hard or fast he will naturally sweat, this undercoat becomes wet and does not dry out readily. Meanwhile the effect is as though one had gotten so overheated that ones clothes were ringing wet, yet were forced to stand still in an unheated apartment without changing. Inasmuch as the horse cannot change his underwear we clip off his heavy coat and give him a warm blanket to make up for it. In this way he does not become overheated as readily, and when he does he dries out quickly.

Clipping is quite a job, especially to the uninitiated. It takes patience, experience and a knowledge of horses. Some few horses will pay no attention to the clippers and stand as steady as rocks, others become quite hysterical. In the latter case two people will be needed.

Modern clippers are simple to use. They consist of a head with two blades which work back and forth on each other, and a cylindrical motor which is held in the hand. They overheat rather readily and the blades become dulled with use when they may be sharpened or replaced. Have handy an open utensil, large enough to take the head of the machine and with kerosene in it to completely cover the blades. Have also a little can of machine oil. Every few minutes put the head and blades into the kerosene, the latter still running, this washes away the loose hairs. Then drop a few drops of oil on the blades and in the holes in the head. These precautions will help to keep your motor cool and your blades sharp. Test the blades with your hand for over-heating, almost any horse will resent clippers that are so hot that they burn!

In clipping it is a good idea to start with the head of the horse as this is the part about which he is most touchy and it is wise to get it over with before the blades get too hot or your animal too restless. If your horse is very nervous about his ears you may have to use a twitch (see page 70), but avoid it if possible, sometimes, especially with young horses being clipped for the first time, you can quiet them just by talking and by doing only a little in one place at a time. Do a patch on the cheek, work up towards the ears and if the horse is nervous go back to the cheek, etc. Clip always against the growth of the hair. Don't press down too hard and on the other hand don't go so lightly that you skip patches. An even, light hand is best. Try not to leave "rail-road tracks" but if you do, don't be too discouraged, they will all disappear in a week or so. You will have to turn the ears wrong side out to get the fuzz inside. Some people don't bother with the ears at all if the horse is restless but a clipped horse with rough ears looks very unfinished. In clipping the front legs stand or crouch a little to one side otherwise a quick movement of the horse's knee will catch you unexpectedly on the chin or forehead. Clip the inside of each leg from the opposite side, reaching under the horse to do so. A restless horse will sometimes hold his feet still if you gently pinch the opposite leg from the one you are working on. If you have a helper have him hold up one leg while you work on the others. Under the throat is a hard place to do neatly as are the brisket, elbow and flanks. This is because the skin is so loose there. For the throat the horse's head must be stretched out and turned away from you. One front leg must be extended straight out its length to make the skin at elbow and brisket taut, this is awkward if you have no helper but can be done by bracing the extended leg with one of your own. At the flanks the skin is pulled away from the clippers with the other hand. A quiet horse can be clipped in from one to two hours, a restless one may take you most of the day. Many people, particularly those who use their horses for hunting or for riding out of doors in rough country prefer to give a "hunting trim" in the fall. This means that they leave their hair long on the legs and under the saddle. The hair on the legs acts as a protection against brambles and that under the saddle is an additional pad to protect the back of horse. Furthermore, the blanket does not protect the legs of the horse standing in the stable. Horses with fine coats that are kept well blanketed will often require only one clipping a year, in the late fall. Ponies always need two, the second coming as soon as the weather settles in the spring and before the summer coat comes in. Ponies have such very heavy coats, regardless of blanketing, that they overheat quickly at the first sign of warm weather. Also the shedding process is extremely disagreeable.

VANNING.

It may at some time be necessary for you to ship your horses from one place to another. You may even be called upon to drive them yourself. Here are a few hints on the subject.

LOADING. Horses, with their inherited fear of stepping on anything unusual, hate to be loaded although in time they become used to it. Many an old race-horse or show horse will climb into the van without urging, turn himself around and back into the stall. Not so with the horse to whom the experience is new or who has acquired the habit of giving trouble. With this type there is always the same routine to be gone through. If you are vanning more than one animal lead a horse that is used to the ordeal in ahead of the greenhorn perhaps the latter will change his mind and follow quietly. If the horse is of the "greedy" variety you may be able to coax him with a pail of oats. Sometimes getting him as close to the tailboard as possible and then lifting each front foot in turn and placing it on the wood will give him confidence.

If all of these methods prove useless there is one remaining one which seldom fails. It needs two people, three are better, and twenty feet of rope. The following is the procedure: The horse is led up to the tail-board or gang plank, one person standing in the van and holding on to his halter shank. One end of the long rope is now attached to one side of the van or to a corner of the tail-board at about the height of the horse's buttock point. The rope is next carried behind the animal, hanging just above his hock and the remaining member or members of the loading crew take up their positions on the opposite side of the tail-board near the van. By steadily pulling on the rope the horse is now hauled, like it or not, into the truck. It is all much easier than it sounds, although the person in the van who is holding on to the halter shank should be ready to jump out of the way quickly inasmuch as when a stubborn animal finally makes up his mind he usually comes in with a bounce.

Regular horse vans or "horse-pullmans" as they are sometimes called, are spacious affairs with built-in stalls and space for the tack. You may not be so fortunate as to be able to carry your animal in one of these, but may be reduced to the lowly cattle truck. This is not so easy, especially if only one animal is being vanned and the truck is large. It may be wise to spend a little money and build a temporary stall along one side so that the horse is not thrown around too much in starting, stopping and turning corners.

In such trucks some drivers like to carry their horses with their heads to the rear, thus putting the weight of the load to the front, others will load them in sardine fashion, head to tail across the truck. In the latter method they keep their balance better in starting and stopping but are unsteady on the curves. If you are doing the driving yourself try and hit an even rate of speed and stick to it. Take the corners easily and shift gears with as little jerking as possible. Pick out your route with regard to smoothness of road and absence of hills.

Make sure that the floor of the van, if at all slippery, is covered with sand or grit to give the horse a footing. Don't feed or water your horse just before vanning but if you want to hang a bag of hay where he can reach it while travelling that is permissible.

Low trailers make excellent vans for one or two horses and one sees every variety from the expensive, shiny, streamlined model to the simplest of home-made affairs. If the coupling is good and the trailer wide enough so that it does not tip readily it is a splendid conveyance and much easier on horse and driver than the high truck.

In cold weather, the horse being vanned will need to be blanketed and perhaps hooded. You may want to bandage his legs using a woollen or cotton bandage with an inside padding of cotton wool, then if he slips he won't bark his shins. If he nips and there is more than one occupant of the van you should muzzle him. If the weather is warm and the van a closed one he will be better off without blanket or sheet. However you take him, you may be sure that when he gets to the end of the journey he will be glad and so will you.

TIEING.

Methods of tieing a horse in his stall are described on page 45, but there are other times and places when it may be necessary to tie a horse and it is just as well to learn the safe way of doing so. First, never tie a horse by the bridle reins if you can help it, even though your horse normally stands quite quietly something may happen to frighten him and then you have a broken bridle to fix. If you must tie him *by the bridle,* find a swinging limb higher than his head and tie to that. Thus, if he pulls back suddenly, the limb will give and no harm will be done, and there is no danger of his getting his reins under his feet. If you think you are going to have to tie your horse outside away from home, take along a halter and halter shank. The halter is put on under the bridle which may have to be let out a bit to accommodate the former, especially if it has a cavesson. One end of the halter-shank which should be equipped with a clip, is fastened to the ring in the halter, the rest is carried as follows: Throw the loose end over the horse's neck just in front of the withers, now catch it on the far side and bring it on top of the rope coming from the halter (see A page 71). Next bend a loop behind where it crosses and fold over (see B). The hanging end is then wrapped around and around the bend until it is all used up, the tip being put through the loop at the end of the bend towards the horse's head. A smart pull on either side of the finished knot tightens it. This is the knot used in the cavalry and called the halter-shank or pigtail knot.

WITHOUT BRIDLE OR HALTER. Sometimes it is necessary to tie a horse when one has only a rope; no bridle and no halter. There is only one safe way to do this. Start by putting a tight knot in one end of your rope. With this end, measure the circumference of the horse's neck near the throat and make a simple knot there, i.e. the first step in a square knot. Do not pull this tight. Put the rope around the animal's neck and run the first knot on the end through the second and tighten up on the latter. This is a knot which cannot slip and choke the horse and which is fairly simple to undo.

FOR MEDICATION. The same method of tieing is used when it is necessary to hold a horse's head up and open its mouth for the purpose of giving him medicine. Here you need a rope long enough

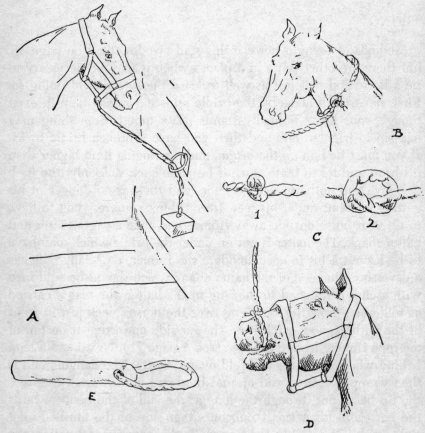

A. A Safe Method of Tieing a Horse in the Stall
B. Tieing a Horse with Neither Bridle Nor Halter
C. This is the knot shown in "B." 1. The knot at the end of the rope is slipped through 2 which is then tightened. The distance between the knots is the circumference of the animal's neck. This same one is used in figure "D" to tie a horse's head up in order to give him medicine.
E. The Twitch is used in quieting restless horses while being treated for injuries. The loop is put around the upper lip and the pole twisted until the pressure is sufficient to distract the horse's attention.

to go from the horse's elevated head over a rafter above him and down to the ground. Use the knot described above but instead of putting it around his neck, wrap it around his upper jaw, running it through his mouth just in front of the grinders. By throwing the other end over the rafter and pulling gently the horse's head is

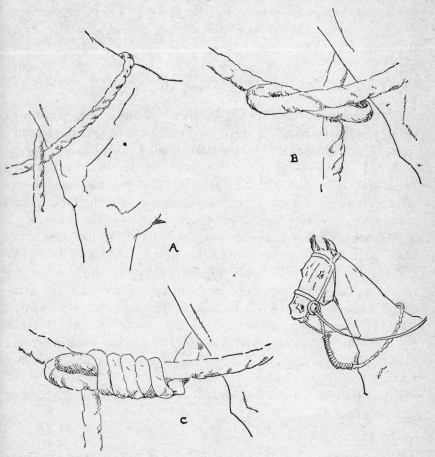

The Cavalry Halter Shank or "pigtail" knot is tied in this manner. The purpose of this knot is to carry a long rope for tieing.

raised into position and he cannot pull away. A word of caution, hold on to the end of the rope, don't tie it down, if the horse becomes too excited you may want to let go of it in a hurry.

WITH HALTER AND SHANK. In tieing a horse to a tie ring, assuming that he wears a halter, run the end of the shank through the ring and then tie the beginning of a slip knot, but instead of pulling the *end* through, make a loop and pull that through, leaving the end dangling. A quick jerk on the loose end will untie the knot. The ranch knot is a variation of this in which the loose end is stuck back through the loop and the whole tightened slightly. This knot is harder for the animal to undo and it will not jam the way a straight slip knot does, although it is not so easy to unfasten in a hurry than the other knot.

COST IN HOURS OF KEEPING A HORSE.

If you are active and quick, whether you are a man, woman or child, you should be able to take good care of your horse and equipment by spending from an hour and a half to two hours a day on him. This is in addition to the time you may want to give to riding or driving him. You will also want to allot one morning or one afternoon a week to extra jobs such as going over the tack or harness for repair, fixing up the stable, clipping, etc. The following is a good schedule for a person who wants to run in the care of his horse with an otherwise active life. Before breakfast, feed your horse and clean up his stall (fifteen to twenty minutes). After breakfast, do your grooming and take care of any first aid treatment. This should take a half hour at the most. When you finish your ride clean your tack and brush your horse off quickly. Take twenty minutes to a half hour for this. Ten or fifteen minutes takes care of the night feeding and rearrangement of bedding. If your horse is fed at noon he will get oats only and five minutes will be the time required. Total time, less than an hour and a half for one or two horses.

CHAPTER V

First Aid

First Aid for horses as for humans is the *immediate and temporary care given in case of accident or sudden illness before the veterinary comes.* Its purposes are as follows: (a) To prevent accidents. (b) To prevent further injury in case of accident. (c) To teach diagnosis of common ailments and injuries so that in calling the veterinary the pertinent symptoms can be described. (d) To teach simple remedies and treatments which may safely be applied if it is not possible to get a veterinary.

IN EVERY INSTANCE, UNLESS THE INJURY IS VERY SUPERFICIAL, CALL A QUALIFIED VETERINARY.

LAMENESS.

Common causes—
  a. Bad shoeing, contracted heels, corns
  b. Bruised frog or heel
  c. Nail in the foot
  d. Gravel
  e. Strained or bruised tendons
  f. Sprained or strained joints: shoulder, knee, fetlock, hip, stifle, gaskin
  g. Laminitis or founder
  h. Thrush
  i. Infections or infected wounds
  j. Splints
  k. Scratches
  l. Rope burns
  m. Wind puffs

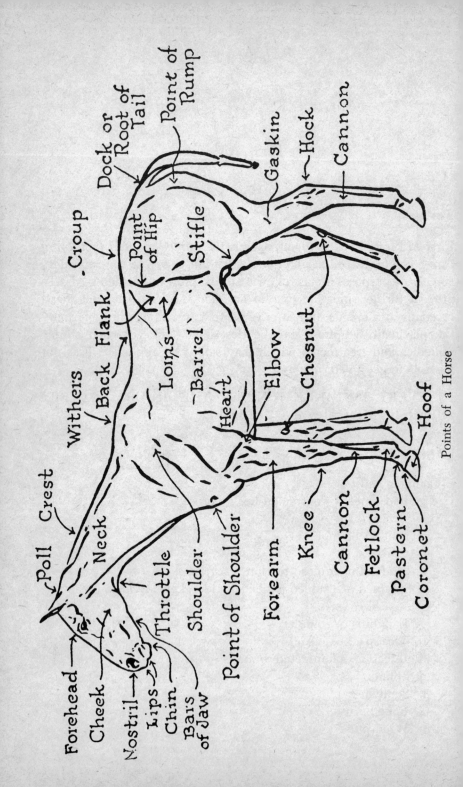

Points of a Horse

PREVENTION OF LAMENESS. Judgment in using your horse and conscienciousness in his care will reduce lameness due to accident or illness to a minimum. A lame horse is useless, a permanently lame horse must be shot. Never gallop your horse over hard or rocky places. Never jump a fence without knowing what the landing is like. In starting new work or in bringing a horse in from pasture, let your early schedule of training be moderate so that his muscles and tendons can become accustomed to the new strain which is being put upon them. Do not stop or turn your horse suddenly unless he has been specially schooled for this. Warm him up before you trot, cool him out before stabling him. Pick up loose wire, ropes, boards (especially those with nails), etc. around stable, paddock and pasture. Tie him properly both in the stall and away from home (see page 69 on tieing). Treat all injuries and ailments immediately and continue the treatment as recommended. Call your veterinary without delay.

DIAGNOSIS OF LAMENESS. Try to locate the seat of the injury as follows: Clean out horse's foot carefully. Tap the sole and frog with a hard instrument, if the horse flinches it indicates soreness at this point caused by a stone bruise, a corn, or a puncture wound due to a nail or other sharp instrument. Examine the frog carefully for symptoms of thrush, i.e. a strong smell and a watery discharge. Feel the wall. Undue heat might indicate a condition of founder or gravel. In feeling for heat on any part of the horse always test by putting the other hand on a corresponding part, you will then be able to tell if the injured part is hotter than it should be. Inasmuch as the condition of founder or laminitis is usually present in both front feet it is wise to use the wall of one of the back feet as a check.

If there is nothing in the foot and no signs of heat or tenderness there test the leg. Starting at the elbow (the stifle if it is the back leg which is lame) run your hand slowly down to the coronet. Symptoms of injury are as follows: (a) Heat, (b) Tenderness, (c) Swelling.

By lifting and bending the leg ascertain if the horse is willing to move it freely. A horse that is lame in the stifle does not like to have his leg lifted high or stretched forward or back. A horse lame in the shoulder often stands with that leg relaxed and only the tip touching the ground. If there is no indication of injury on the leg itself go to the higher joints, shoulder or hip. Press and pinch these joints to test for tenderness. A horse lame in the shoulder does not like to back, particularly if in doing so he has to raise the injured leg. The old test used to be to back him over a log of wood. Another test for shoulder lameness is to turn the horse very short first in one direction and then in the other. There will be exaggerated lameness and stiffness on the side of the injury.

BOWED TENDONS. Tendons are the stringy cords by which the muscle is attached to the bone to give movement. These cords are easily strained or whipped when the horse puts a too great or an unaccustomed tension on them. The tendons then swell and become hot and inflamed. Those directly behind the cannon bones take the

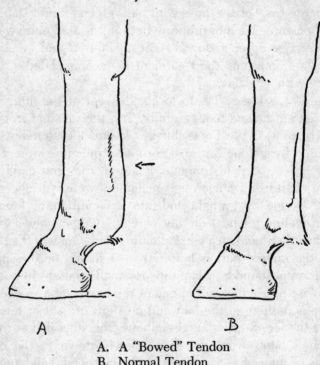

A. A "Bowed" Tendon
B. Normal Tendon

form of a bow, hence the name. The horse is usually very lame. This lameness may or may not be permanent, the enlargement usually remains for life. The veterinary will usually recommend blistering or, in extreme cases, pin firing. In calling him be sure and give the symptoms so that he can bring the necessary materials with him. The horse must have complete rest after the treatment. He should have his shoes removed and be turned out to pasture for some months.

CURBY HOCKS. A curby hock is recognized by a swelling directly below the point of the hock. Blistering is usually recommended. In mild cases it may only be necessary to apply Churchill's Iodine to form a very light blister. Paint the affected part once daily for three days and then stop treatment for three days. Repeat until the skin

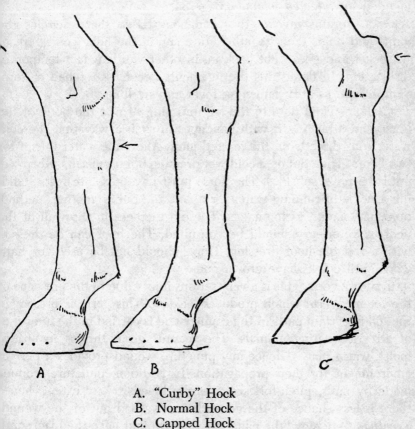

A. "Curby" Hock
B. Normal Hock
C. Capped Hock

becomes slightly blistered. If the horse is not lame he may be used. In severe cases he may need a heavy blister and complete rest.

CAPPED HOCK. This is an injury to the membranous covering of the hock. Treat as recommended for *curby hocks*. In mild cases the condition can sometimes be helped by cold applications of epsom salts, see page 78, *Sprains*.

GRAVEL. Sometimes a little piece of gravel or sand will work itsel
in behind the wall and travel upwards through the hoof, coming
out eventually at the coronet. Early symptoms are (a) Lamenes
may be severe or only intermittent, (b) Heat in whole area of wall
Later symptoms include a pimple or boil at the top of the corone
which will open and exude pus. The gravel can be felt and ex
tracted. *Treatment.* Treat with hot epsom salts compresses as recom
mended for *infected wounds,* page 84.

SPRAINS. Sprains involve the tendenous tissue that encircles the
joints and holds them together. In a sprain the bone comes out o
its socket and goes right back in, in other words it is a temporary
dislocation. In doing this the ligaments become stretched or torn
Symptoms are—great lameness, fever and swelling.

Treatment—Tie a hose to the leg, bringing it over the body of the
horse and attaching it with bandages in such a way that the cold
water runs directly on the injured joint. Allow the water to run fo.
two hours. Then put on a cold wet dressing of epsom salts. Keep we
until the fever subsides. The horse must have absolute rest. Stand
ing a horse in running water for about an hour is another method

QUARTER CRACKS, SAND CRACKS. These are cracks in the wall of the
hoof. A veterinary should be consulted. The hoof can be dressed
with a reliable hoof dressing. Horse should not be used for hard
work until cracks have grown out.

NAIL IN THE FOOT. This is a very serious injury due to the presence o
the tetanus germ which produces lockjaw. This germ is present in
the soil in certain parts of the country and is carried in the intestine:
of all grass eating animals. The tetanus germ thrives in airless
moist, warm places, hence any puncture wound provides a perfec
environment for their propagation. In addition, puncture wound.
made by nails, pitch-forks, etc. rarely bleed to any extent, hence
there is less chance of the germs being washed out of the wound
Treatment—Remove the nails, if a deep wound have the blacksmith
pare it out. Soak the foot in a hot lysol solution. Work Churchill'
iodine well into the wound. Consult your veterinary as to the advis
ability of having the horse given an injection of anti-tetanus serum
This will depend on the gravity of the wound and the prevalenc
of tetanus in your vicinity. Remember that lockjaw, once contracted
is fatal. The injection is a preventive.

SHOE BOIL. Sometimes called capped elbow. This is a boil which orms on the point of the elbow. It may be as big as an orange and usually appears rather suddenly. The horse is lame inasmuch as when he moves the boil is pressed against his side.

Cause—Some horses have a habit of sleeping with their feet bent under them or of pressing the elbow against the shoe in rising.

Treatment—When the boil comes to a head the veterinary will open it allowing the pus to escape. *Do not squeeze* any type of boil. Nature provides a thickening of the tissues behind an injury of this sort which prevents the infection from spreading. By squeezing you may injure this protective tissue, permitting the escape of the infection into the blood stream or the lymphatic system. Syringe inside the enlargement using a small, rubber ear syringe and a mild boric acid or lysol solution. Do this twice daily until all discharge has stopped. Work the syringe freely around the opening of the wound so that this will not close up until the infection has been healed from within. Cover surrounding hair especially below the wound and well down the leg with a protecting film of vaseline. This will prevent the pus from sticking to the hair and will help to prevent the latter from falling out. Keep the whole area very clean, washing all over with the lysol or boric acid solution. If the lump remains after the discharge has stopped it may be painted on the outside with iodine as recommended for *curby hocks,* see page 77.

Shoe Boil Boot

Or it may be rubbed every day with iodex. As soon as the boil appears the horse should have his shoe removed and a specially made shoe which will not touch the part should be put on in its place. He should also wear a "shoe-boil boot," see illustration. Such a boot may be improvised out of a piece of inner tube and some sand. If the horse is stabled in a straight stall a piece of two by four nailed across the floor of the stall three feet from the front will tend to make him spread out his legs and so prevent a recurrence of the boil.

SHOULDER LAMENESS.

Symptoms—see page 75 *Diagnosis of lameness.*

Treatment—In mild cases rubbing the shoulder muscles with any good liniment or with a solution of olive-oil and turpentine will help. Blistering with iodine is used for more severe cases. In serious cases a standard spanish fly blister may be required. The veterinary must put this on as he can tell by the condition of the horse how severe a blister is needed and will apply it accordingly. The more a spanish fly blister is rubbed in and the longer it is left on before being washed off the more severe it is. Whenever such a blister is applied the horse's head must be tied in such a manner that he cannot reach the blister with his muzzle, and kept so tied until the medication is removed.

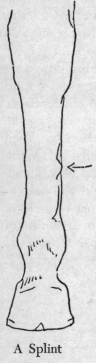

A Splint

SPLINTS. Splints are hard, horny excrescences which form on the inside of the cannon bones. Common in young horses, they seldom cause permanent lameness. If large and placed low on the bone the horse may hit them inadvertently with the side or heel of the opposite leg causing him to go dead lame for a few steps from the pain of the blow. If such is the case a shin boot should be worn.

Cause of splints—Hard or fast work when the bones are still soft or the horse not in condition. Too hard work on hard roads. Constant jarring as sometimes happens when a horse has the habit of pawing in his stall.

Treatment—It is usually necessary to apply a blister when the splint forms. In severe cases the part may have to be pin-fired.

LAMENESS IN THE STIFLE JOINT. The stifle joint is one very prone to injury, especially to dislocation. In the latter case the veterinary will reduce the dislocation and probably follow with a blister to tighten up the ligaments.

LAMENESS DUE TO WOUNDS OR INFECTIONS. See page 84, treatment of infections.

LAMENESS DUE TO CHRONIC CONDITION OR DISEASE.

FOUNDER OR LAMINITIS.

Cause—If a horse is worked until he is exhausted or covered with lather and is then allowed to drink his fill of water and stand, it affects the blood vessels in his feet. The walls of the capillaries (the small blood vessels through which the blood passes from the arteries into the veins to be returned to the heart) are permanently injured. The coffin joint, inside the hoof, is injured and drops out of position. The sole also drops.

Symptoms—Severe lameness in both feet. The horse tips along as though walking on eggs. Excessive heat in entire foot and wall. Founder is usually present in both front feet. This fact helps to distinguish it from other types of lameness due to an injury or gravel, etc.

Treatment—Mild cases may be relieved by letting the horse stand in a stream of running water for several hours. Standing in deep mud may also help. The horse must have his shoes removed but the blacksmith must be very gentle in so doing as the condition is extremely painful. The veterinary may prescribe a "founder medicine." If the horse recovers to the extent that he is lame only until his blood is warmed up he may benefit by wearing rubber pads. But he will never be up to any except very light work.

SCRATCHES. This is a condition similar to chapped hands in people. It is caused by cold, wet mud, consequently is most prevalent in the spring of the year when horses are first turned out to pasture. It can be prevented by thoroughly cleaning and drying the backs of the pasterns each time the horse is brought in.

Symptoms—Raw, scabby wounds sometimes with a watery discharge located at the back of the pastern just above the heel. Usually present in both front or both hind feet.

Treatment—Soak the feet in warm water containing boric acid, lysol or epsom salts. When the parts are softened, pick off all scabs and crusts. Dry carefully, cover thickly with B.F.I. powder (Bismuth, Formic Iodide). Keep the horse off muddy ground until entirely healed. If the scabs reoccur the treatment must be repeated but as a rule the condition will disappear in a day or so if all crusts have been removed and the parts kept well dusted.

THRUSH. This is a contagious and very common disease. It appears

most unexpectedly for no apparant reason and disappears just as
quickly.

Symptoms—Extreme and very sudden lameness. On examination of
the foot the frog is found to be soft and spreading with a discharge
of watery fluid and a very strong odour.

Treatment—There are any number of remedies for thrush, every
blacksmith and veterinary has his favourite. Blue stone (copper
sulphate or blue vitriol) is a very satisfactory one. This must be
crushed fine and a few pinches worked well into the frog and kept
there by means of a packing of cotton. The horse will often recover
after one application.

WOUNDS.

A wound is a break in the skin or mucous membrane covering the
body. There are four types of wounds as follows:

Puncture

Abrasive

Incised

Lacerated

Remember them by the key word, P A I L . Wounds are dangerous from the point of view of infection and from the point of view
of bleeding.

PUNCTURE WOUNDS. Puncture wounds are those caused by any sharp
instrument such as a nail, a pitch fork or the spoke of a wheel. The
danger and treatment of puncture wounds are discussed on page
84 under "Nail in the foot."

ABRASIVE WOUNDS. Abrasive wounds are those in which the surface
of the skin has been rubbed off due to friction. If located on the
body of the horse they usually heal very readily and seldom leave
a scar. All types of wounds heal more slowly when located on the
legs of a horse due to the scanty amount of flesh and the fact that
each leg is fed by only one main artery, thus the circulation of the
blood is not as active in these parts.

Treatment—Paint the parts with methylene blue. This may be
bought in a suitable solution for treatment of horses in many feed
and horse equipment places. Saddle galls, collar and girth gall
yield readily to this treatment. If the injury is due to one of the
above. the horse must not be used until the place has entirely

healed and the tack should then be padded or changed so that it
does not rub him again.

ROPE BURNS. They are abrasive wounds caused by the horse getting
entangled in a rope. They occur most commonly on the back pastern
and are difficult to heal in this place because of the motion of the
joint. If the wound does not yield to treatment suggested above,
put a thick coating of *resinol ointment* on it. This ointment allows
the wound to heal without forming a scab. B.F.I. powder is also
effective in some cases.

INCISED WOUNDS. Incised wounds are those in which the tissues are
cut cleanly as with a knife. They are not as common as other types
of wounds. If deep there may be danger due to excessive bleeding,
(see control of bleeding, below). If extensive they may have to
be stitched. Incised wounds are less subject to infection than other
types of wounds. They should be painted with iodine or methylene
blue and kept clean.

LACERATED WOUNDS. Lacerated wounds are those in which the tissues are torn. They infect very easily, heal slowly, especially on the
legs. If extensive they will have to be stitched. Certain parts of the
body of the horse such as the shoulder point and the flank, do not
retain stitches due to the looseness of the skin and the movement. It
is sometimes possible to keep the flap in place by means of broad
bands of adhesive tape. These will have to be replaced each day,
they should extend from six to ten inches beyond the wound on all
sides and the hair to which they are attached must be perfectly dry.

Lacerations which are slight may be painted with methylene
blue or iodine and powdered with B.F.I. Serious lacerations almost
always become infected and should be treated accordingly, see
page 84.

CONTROL OF BLEEDING. The blood is pumped by the heart throughout the body running through the arteries. These have thick walls
and are deeply imbedded. Consequently they are rarely severed.
They branch off and become smaller and smaller ending, when they
reach the surface, in little blood vessels known as "capillaries." The
blood passes through the walls of the capillaries returning to the
heart by way of the veins. On the way out it carries oxygen and
food to the tissues, on the way back it carries the refuse and the
carbon dioxide, the former being carried off by way of the kidneys
and bowels, the latter being exchanged for more oxygen in the

lungs. The pressure is higher in the arteries than in the veins, so when an artery is injured it is easily diagnosed by the spurting. Blood from veins flows, blood from capillaries oozes. Except in such accidents as those occuring when a horse is being vanned, arterial bleeding is rare. This is fortunate as it is hard to control and usually fatal. If there is spurting from a leg, apply a tight tourniquet around the leg above the wound. Release this every fifteen minutes or the horse may develop gangrene. If an internal artery is cut there is little that you can do except apply pressure between the wound and the heart.

Venous bleeding is controlled by direct pressure on the wound itself, holding a pad against the part until a scab has formed. Capillary bleeding stops of itself in a few minutes.

INFECTIONS AND INFECTED WOUNDS.

Cause—Germs are present everywhere. When the covering to the body is penetrated they enter the wound and instantly begin to propagate and to become more virulent. As soon as this occurs nature sends the soldiers of the body, the white corpuscles and the corpuscles of the lymphatic system to fight off the invasion of the enemy. She also surrounds the battlefield with thickened tissue to prevent the spread of the infection. During the battle between the corpuscles of the body and the germs many or both are killed. This is then exuded in the form of pus. It is very important that the wound be kept open until the enemy is completely conquered so that all this dead matter can be gotten rid of.

Early Symptoms of Infection—Excessive heat, swelling and soreness.

Later Symptoms—Discharge of pus if the infection occurs in a wound. Signs of the swelling coming to a head if the infection occurs where a wound has already healed or if it is a form of boil.

Treatment of Infections—Prepare and have ready the following: A pail of hot water, the temperature should be such that you can hardly bear your hand in it, but not so hot that it will burn. Into this put a handful of epsom salts. You will also need a turkish towel, a roll of absorbent cotton and a horsebandage (see page 90). If the horse is restless you will need a helper but often getting him used to the heat very gradually and soothing him by talking to him will suffice.

Dip the towel in the solution and wrap it around the infected area. Hold it in place until it has cooled slightly. Repeat this treatment for twenty minutes. If the water becomes cold more hot water should be added.

Next soak the absorbent cotton in the hot epsom salts solution, wrap it around the parts and bandage in place with the bandage. The bandage may be put on in the form of a spiral, with a half twist each time it goes around the back of the leg to take up the slack, or, if the injury is at a joint, it may be put on in the form of a figure eight.

If the injury is at some point where it cannot be bandaged it may be well to follow the bathing with epsom salts with a poultice of either flax seed or antiphlogistine. The former is made by pouring a little boiling water on flax seed and allowing it to cool just sufficiently so that it will not burn. This is often used on the sole of the foot where it is kept in place with a layer of gauze on top of which is placed several thicknesses of newspaper shaped to fit the inside of the shoe.

Antiphlogistine is prepared as follows:

Put the can into a pan of boiling water having first removed the metal cap. The water should be three quarters of the way up the can. Stir with a knife blade so that the heat will be distributed evenly. To test, hold the knife blade with a little of the paste on the back of your hand, it should be very hot but not hot enough to burn.

Leave the poultice on for thirty-six hours. If hot wet dressings are being used they should stay on for a similar length of time and be constantly soaked with the epsom salts solution to keep them wet and hot.

On removing, the infection will either have subsided, or it will have opened or come to a head and be ready for opening.

Once the pus has begun to be exuded the following is the treatment: Bathe the parts with a good antiseptic solution such as lysol. If the wound is deep, syringe out thoroughly using either a rubber ear syringe or an enema bag. Balsam of Peru, and B.F.I. powder, are both excellent as healing agents and to keep off flies. Sulphathiazole is excellent for infections. Keep all surrounding areas well protected with a coating of vaseline.

FISTULA OF POLL EVIL. This is a form of boil which forms either on the withers or on the poll. Fistula of the withers is caused by an ill fitting saddle and should not be tolerated in any well run stable. Poll Evil can come from a horse rearing or throwing his head up and hitting it on a beam. Treatment is the same as for *treatment of infections*, see page 84.

COMMON ILLNESSES.

COLIC. A horse has no vomiting muscles, therefore, once something enters his stomach he cannot eject it except by way of the intestines, kidneys and bowels. Colic, which is really acute indigestion, is the indirect cause of many deaths. In flatulent colic the horse generates a great deal of gas, this gas may press on the heart to such an extent that death ensues. The veterinary endeavours to stimulate the heart and to relieve this pressure on it.

Causes—Incorrect or excessive feeding. Watering directly after a full meal of grain. Eating fermented grass (grass that has been cut but has not been properly cured). Insufficient mastication of food which may be due to injury or condition of teeth.

Symptoms—The horse appears very uncomfortable. He bites at his side, gets up and lies down repeatedly. He may break out in a sweat. In flatulant colic his sides swell and become stiff, the horse may roll over on his back and kick or extend his legs.

Treatment—Take the horse into an open space where he can have some freedom. Cover him with a blanket. Rub his loins with a liniment. If you have a colic remedy on hand give it to him. CALL THE VETERINARY AT ONCE. While you are waiting for him to come give the animal an enema of soda and water to help him expel the gas. Fill the enema bag several times. COLIC IS SERIOUS, OFTEN FATAL, DON'T DELAY. When the veterinary cames he may have to cut a slit in the flanks of the horse to allow the gas to escape. He will probably give him a heart stimulant. Extract of Nux Vomica is often used, it is very strong and the dose is a teaspoonful on the tongue.

FODDER POISONING. There are many weeds which poison horses. Due to his inability to vomit these may cause death.

Symptoms—The horse goes off his feed. He appears restless and hangs his head. After a short space of time he will develop a weakness in his hind quarters, being unable to stand up. CALL A VETERINARY AS SOON AS THE FIRST SYMPTOMS ARE NOTICED. He may be able to save the horse with medication and heart stimulants.

FOUNDER. See page 81.

COUGHS AND COLD.

Cause—Exposure, overwork, bad feeding. Colds are contagious and readily passed from one horse to another especially via the common watering trough.

Symptoms—A hacking cough. A discharge at the nostrils. The horse may go off his feed and run a high temperature.

Treatment—In all cases call a veterinary. Keep the horse well blanketed, warm and out of drafts. If he will eat give him a hot bran mash. The veterinary will prescribe medication. Oil of tar and aconite is an old remedy for coughs.

SHIPPING FEVER. Common in young horses shipped in from the West. It is very contagious. The symptoms are the same as for colds.

Prevention—All new comers to the stable should be isolated for a few days.

Treatment—Call the veterinary. The sulpha drugs are often used for this disease, your doctor will have to prescribe as they are not for sale without a prescription.

STRANGLES. Shipping fever sometimes develops into a condition known as strangles. There is a tremendous discharge from the nose. The glands around the throat and under the jaw swell and the veterinary will open the latter. This is then treated like a boil, see page 84, infections.

THRUSH. See page 81.

HEAVES. Heaves or broken wind is a disease of the respiratory system which is also associated with the stomach. Its symptoms are similar to asthma in humans. A hacking or a deep, hollow cough. Great difficulty in breathing and shortness of wind. The horse has a peculiar double beat in expiration. The flanks of a horse with heaves will heave in and out after only mild exertion. In light cases the hacking cough disappears when the horse is warmed up to his work but returns as soon as he is exposed to dust. In serious cases the cough is deep and very loud. This disease is generally associated with old and overweight horses.

Causes—Overwork, bad feeding, dusty hay, etc. are some of the causes of heaves.

Treatment—There is no cure for a bad case of heaves although some of the symptoms can be alleviated as follows: The horse should have only slow, light work. All feed, both grain and hay, should be wetted down before feeding. Avoid riding on dusty roads. Use a recommended cough medicine.

INFLAMMATION OF THE KIDNEYS. The horse's kidneys are situated in a very vulnerable spot being in the small of the back near the surface. They can be injured by a blow or by a weight on the loins such as occurs when a rider sits too far back. They can also be injured by exposure to cold.

Symptoms—Great tenderness and soreness in the regions of the loins. The horse flinches when groomed and sags when mounted. The urine is scanty, thick and discoloured. The horse stands in a "stretching" position and spraddles when he walks.

Treatment—There are a number of kidney remedies on the market. Sweet spirits of nitre, a tablespoonful or two in a little water is often prescribed. You can easily diagnose this disease and should consult your veterinary at once as in so doing you will save your horse a great deal of pain.

LAMPERS. This is a condition of the blood causing swelling of the roof of the mouth and the gums. The old treatment used to be to burn or lance the parts but this is no longer recommended. Give your horse a good dose of Glauber salts, a big handful in his feed. Give him soft food for a few days until the inflammation disappears. A similar condition of the blood can cause "stocking up." The back legs swell when the horse stands for any period of time. In addition to the salts the affected parts can be rubbed, rubbing towards the heart, and then bandaged.

CARE OF THE TEETH. The horse's back teeth are called grinders and act as mill stones to grind up the food. From this motion they sometimes develop sharp edges and points. Unless these are periodically filed away they will prevent the horse from chewing properly and in so doing will cause him to lose weight, get out of condition and may bring on colic. This process is called "floating" the teeth and is not painful. Have your veterinary inspect your horse's teeth every six months to be sure that they do not need this attention.

WOLF TEETH. Wolf teeth are small extra teeth which sometimes appear just in front of the foremost grinders. They are about the size of your fingernail and are said to make the horse dopey. It is also contended that the wolf teeth are situated near the nerves of the eye, and that the action of the bit may affect the horse's sight. They are easy to remove and should be removed as soon as discovered.

OPTHALMIA. A very common and serious disease of the eyes. It is also contagious and some forms are recurrent, going from one eye to another, disappearing for a time and then returning.

Symptoms—The lid is swollen and closes. The eye waters profusely. The haws become reddened and protrude. The eyeball has a bluish caste. In many cases the horse loses his vision.

Treatment—Bathe with boric acid solution. Call a veterinary.

INFLAMMATION OF THE EYES. This is often due to pollen, dust or hay seeds. It is common in the fall when the rag weed abounds. Horses fed hay from racks high over their heads are subject to this.

Symptoms—Watering of the eyes, some inflammation of the haws.

Treatment—Bathe with warm solutions of boric acid or epsom salts.

CATARACT. A membraneous growth on the eye which can be removed surgically.

BLOWS TO THE EYES. These often cause blindness and every care should be taken to prevent such injuries. When one eye is blinded a sympathetic blindness may appear in the other. Veterinaries sometimes remove the blinded eye to prevent this condition. Horses with only one eye soon become accustomed to it and can often carry on as well as before. Treat any injuries to the eyes with cool compresses and call the veterinary.

WALL EYES. Wall eyes are not a disease, but they account for a good deal of the nervousness and headshyness in horses. In this condition one or both eyes will magnify objects, making the horse afraid of the most familiar things when they are waved near his head. If your horse seems more headshy on one side than on the other he is probably wall eyed on that side. Every care should be taken in handling him to prevent his becoming more nervous.

WORMS. Horses are subject to various kinds of worms. If you see signs of worms in the manure report them to your veterinary and he will handle them according to what they are. Many people give small doses of ferrous sulphate (sulphate of iron) at regular intervals to combat worms. The dose is a tablespoonful in the feed once a day for a week. A salt enema for pin-worms is often recommended.

LICE. Horses in poor condition or horses that are not being worked sufficiently to cause them to sweat are often attacked by lice. These lice do not bother humans and seldom pass to other horses. Clip the horse and wash all over with a solution of lysol. Rub kerosene in the mane and tail. Repeat treatment in three days.

FOR THE MEDICINE CHEST.

BANDAGES. Make them of canton flannel, six inches wide, nine feet long. Sew two tapes at one end or use safety pins. Keep rolled up, taped end inside, ready for use.

GAUZE SQUARES. Three or four inch squares in individual wrappings. Use in cleaning and dressing wounds.

STERILE ABSORBENT COTTON. For use in wet dressings, to make swabs for cleaning wounds, for washing eyes. Never put absorbent cotton against open wound as a dressing as it adheres to the tissues.

METAL SYRINGE. For giving medicine internally.

RUBBER SYRINGES. Ear syringe for treating wounds and boils, enema bag for use in colic and intestinal inflammation.

VETERINARY VASELINE. Use around wounds to keep clean and to prevent loss of hair.

BORIC ACID SOLUTION. Use in eyes.

EPSOM SALTS. Use for sprains, strains and infections.

BALSAM OF PERU. Very healing, for bad wounds.

SULPHATHIAZOLE OINTMENT. For infected wounds.

CHURCHILL'S IODINE. A strong solution of iodine. Use in puncture wounds and as a mild blister.

METHYLENE BLUE. Disinfectant for abrasions and lacerations.

GLAUBER SALTS. A purgative, dose, a handful in the feed.

BLUE STONE. (copper sulphate, blue vitriol)—Used for thrush.

FERROUS SULPHATE. (sulphate of iron)—Used for worms.

WHITE ROCK. A kind of clay used to reduce inflammation and to pack a horse's hoof if the sole becomes hard.

You will also need a good colic remedy, a cough remedy and a kidney remedy.

CHAPTER VI

Handling the Horse

An old saying has it that the only safe horse is a dead one. There is a good deal of truth in this in that horses by nature are timid and nervous and must be watched. Learning a few of the fundamentals about the dispositions and temperaments of horses will make the handling of them both easier and safer.

The first and most important thing to do is to study your horse as an individual. Learn all you can about what he likes and what he doesn't like, what he is afraid of, his disposition and his habits, good and bad. Each animal is different in character from every other, none is perfect, only a few are really ill-intentioned, you had best leave the latter class to the professional.

In addition to the individual characteristics of horses there are certain fundamental ones which have to be taken into consideration. There is their extreme and unreasoning timidity. The horse is one of the few animals which nature neglected as far as defensive weapons are concerned. The cow, the deer and the goat have their horns, the cat family has its claws and fangs, the snake has its poison or its constrictor characteristics. But the horse was originally endowed only with the power of running. It was only after centuries had passed that the little animal's soft toes were changed to the hard hoof of the horse of today. Is it to be wondered that the first reaction of a horse is that of fear, and that he is apt to defend himself with his heels rather than wait to find out whether or not he has any real reason to be afraid?

Because of this reaction, horsemen who want to get the best results make it a practice never to do anything suddenly. Never raise your hands suddenly, never wave a stick or throw a ball around your horse. Never slam a door nor let a gate swing to in his face. Most accidents are caused by neglecting to take into consideration this fear in horses, either on the part of the rider or on the part of the bystander.

The second fundamental trait in horses, and the thing which makes it possible for us to control them, is the fact that a horse can think of only one thing at a time. He is not the intelligent animal that the writers of horse books would have us believe. This is certainly just as well, if the horse, with his speed and his strength, had also the power of reasoning it would be we who were between the shafts!

It takes very little to keep a horse's attention. A favourite trick of the horsebreaker who is called upon to handle vicious animals, is to attach a loose, dangling chain in such a fashion that it jingles and taps the horse on the knee as he walks. The latter, his interest being in the chain, forgets his animosity towards the man. Firm use of rider's legs will help the timid horse to forget his fear.

A horse, with his eyes set as they are on the side of his head, can see as well behind him as before him; a fact which many people forget. Stand directly behind your horse's tail, ten feet to the rear and wave a pitchfork, the horse will notice it immediately and be just as frightened as though you were standing ten feet away but in front of him. Not being able to reason or to deduce, to a horse, a ball that is thrown in the air becomes a tall monster; a dog, coming out suddenly over a wall takes his size from his distance from the ground.

The unaccustomed and the sudden, though it may be no more than a bird flying up suddenly or a bit of paper blown by the wind, is more apt to cause fear than a noisy truck. The latter, to the present day horse, is only one of a class of equally loud and large objects to which he has been accustomed from birth. Horsemen who understand this will be constantly on the alert for the thing which may cause fear. They will see the clothes blowing on the line before the horse does and act accordingly.

Horses learn by repetition, they obey because their muscular reactions are trained to follow certain established patterns when given a definite, established signal or command. The first instinct of the horse is to give way and to obey, the second is to rebel. The trainer spends many hours cultivating the first instinct—subjugating the second. That the horse's every movement can be brought under control is evidenced by the *dressage* horse. These animals are so schooled that they perform the most intricate movements when so commanded by the almost imperceptible signals given by the rider. They are taught to trot and gallop on the diagonal, in place and even backwards; to change leads on designated steps, etc. and to do all these difficult things so fluidly and willingly that they seem to be one with the man. Yet all the time it is the rider who is doing the thinking, giving precise, exact commands for each movement, telling the horse through his hands, his legs and the way he throws his weight exactly what is expected of him.

So much for the fundamentals, now for a few specific directions for handling your horse as you work with him around the stable.

ENTERING A STALL.

It must be remembered that horses doze on their feet, so before entering a stall you should speak to your horse quietly. Next lay your hand on his rump on the left side, causing him to step to the right. These directions apply only to the horse in the straight stall. In the box stall your horse, unless he is eating or unless he is very nervous, will probably come up to the door when you open it. Make sure, in the latter case, that you push his head aside as you enter or he may try to bolt past you. Enter the stall, box or straight, quietly and go directly to the animal's head.

BACKING A HORSE OUT OF A STRAIGHT STALL.

Decide before you begin which way you want the horse to go after he is out of the stall and turn his head slightly in that direction. Now place your left hand on his halter or muzzle, the right on his left shoulder and push steadily continuing to turn his head as you do so. The horse will back out headed correctly. Occasionally one runs across the animal that has a dislike for backing or that has learned that he can cause trouble by refusing to do so. Push a little harder and if necessary give him a sharp kick, first on one cannon bone and then on the other. Reward him with a pat or a bit of carrot when he obeys willingly and he will soon learn. Above all, never give in until the battle is won—by you.

RETURNING A HORSE TO A STALL.

Many people make the mistake of leading the horse to the entrance of the stall and then turning him loose. This is wrong for several reasons. He may push past you suddenly, stepping on your feet or knocking you against the post in his desire to get in to his water and feed. If he has a bridle on he may duck his head and step on the reins. If the grain is already in the manger he will cer-

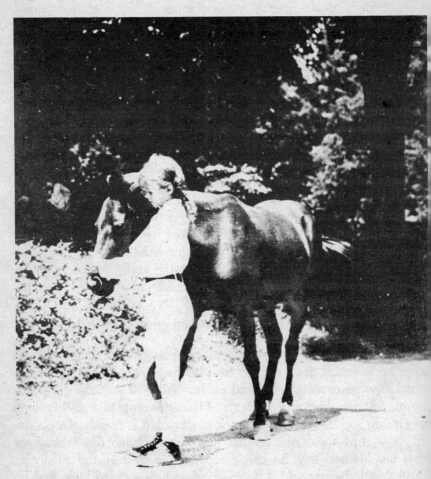

Leading Without Halter or Bridle

In leading this way the right hand leads and the left hand on the nose controls.

tainly get a hasty mouthful which means more trouble in cleaning the bit. He may even dodge past you *and* the stall and out of the stable door to freedom, saddle, bridle and all! The proper way is to precede the horse into his stall, making him follow you in an orderly fashion, and insisting that he wait before reaching for the manger while you remove his tack and adjust the halter.

LEADING A HORSE.

A. IN A BRIDLE. Pull the reins over his head, take the end in your left hand to keep it from dragging, place your right hand on the reins six inches from the bit, grasping all the reins in your hand at his point. Walk straight forward *without looking back at your horse*. In looking back you will unconsciously pull back on the reins and the horse will stop.

B. IN A HALTER WITH A SHANK. Hold the rope as directed above, do not lead by holding only the end, it is easy for the horse to pull away if you do this.

C. WITH NEITHER BRIDLE NOR HALTER. Grasp the horse's forelock in the right hand and put your left hand on his muzzle over the nostrils. It is the latter which will give control if he tries to pull back or jerk away. Slightly pinching the nose just above the nostrils will distract his attention. With a mean, stubborn animal you may have to pinch hard enough to close the air passage. A horse cannot breathe through his mouth and will soon give up if his wind is cut off.

D. WITH NO FORELOCK. Slide a rope or strap (a belt will do in an emergency) around his neck just behind the ears and grasp the two ends at the throat. Never try to lead a horse by the ear, he will only resist.

E. LEADING A HORSE THAT BALKS. If your horse plants his front feet and stubbornly refuses to budge try turning him slightly first to one side and then to the other. Thinking about turning he will forget about not going forward and will usually step out readily enough. Another example of the one track mind!

F. LEADING A HORSE THROUGH A NARROW DOORWAY OR OVER ROUGH GROUND. Face the horse and take a rein in each hand, shoulder high. Walk backwards yourself, thus you can check or steer the horse so that he does not hurt himself. If he is afraid, as many horses are, of bad ground, have a second person urge him gently from the rear while you control him from the front.

Leading a Horse Over Rough Ground or Through a Narrow Barway

WORKING AROUND A KICKER.

A few horses will deliberately kick a man even when not frightened. These animals must be watched very carefully for they will wait until your back is turned and then take advantage of you. Never stand directly behind such a horse or even behind and to one side where he might make a sharp swing and strike you. Be especially quiet with such animals but show no signs of fear. When passing from one side to the other go by way of the head (this is a good rule even when working around a gentle horse). It is sometimes necessary to go around the hindquarters. In such a case either pass well out of range or, starting at the flank, speak to the horse and cause him to move over a step, now pass around his rump, keeping your hand on his quarters and pressing *close* to him. Don't move hurriedly. If the horse kicks when you are in this position all he can do is shove you with his hocks, whereas if he kicks when you are two feet or so behind him he will get you with the full force of the blow. Furthermore, by leaning against him and not hurrying, you show him that you are not afraid of him and in so doing you win half the battle.

Mares in heat will often be more nervous at this time and more inclined to kick. Be careful, too, during the fly season—a horse is constantly kicking at the flies and switching his tail, it is up to you to keep out of the way.

CATCHING HORSES IN PASTURE.

Unless a horse has been badly treated it is easy to teach him to come up to you readily. Carry a carrot or a few grains of oats with you always and whenever you walk into the pasture, call your horse with some distinctive whistle, give him a tidbit, pat him and let him go his way. If he has been handled so little that he will not come up to you, or is man-shy from mistreatment, fix a box in the corner of the paddock and put your carrot in that, standing a few steps away while your horse eats it, and talking to him quietly. He will come to look for the treat and will gradually let you come up to him and stroke him while he is eating. Before long he will be nuzzling in your pocket for more.

One sure way of *not* catching a horse is to chase him, either on foot or from another horse. He can both outdodge you and outrun you and you will only succeed in making a bad habit worse. Above all, never try to corner him and come in from the rear, you are only inviting trouble.

If you have to catch a horse in a hurry and he will not come up to you voluntarily, try the following methods: Start by bringing any other loose animals into the stable, your horse may follow along as they hate to be left alone. If this doesn't work, go out with a pail of oats and attract his attention by rattling the pail. Put a few grains of oats on the grass and walk away. When he has eaten these, put down some more fifty feet or so nearer the stable. Nine times out of ten he will then follow you into the stall; don't try to snatch at him as he eats for if you fail your job will be all the harder.

When even this method won't work the following, which needs two people and twenty feet of rope, will. Coax or manoeuvre the animal into a corner of the paddock. One person now takes each end of the rope and bars off the way of escape, coming gradually towards each other until the animal is penned in the corner and held there, the rope along his side or just above the hocks. All this must be done as carefully and with as little movement as possible for an animal that requires this treatment must be very nervous indeed and you want to avoid making him more so.

BRIDLING.

The knack of bridling a restless horse or one who sets up a defence against taking the bit, is one that looks and is simple—once you have learned how to do it. Until you have learned it seems next to impossible. A beginner can take twenty minutes to put on the bridle that the experienced horseman gets on in twenty seconds.

There are several distinct steps in bridling a horse. The first is the approach. The horseman comes up to the head from behind or from the side and on the left. He carries the crown-piece in his left hand, the ends of the reins in his right. He stops at the horse's shoulder facing front. Never try to bridle a horse while standing directly in front and facing him, he will only back away from you.

The next step is to slip the reins over his head, allowing them to rest just back of the poll close to the ears. If the reins are slipped back on the neck as far as the withers, and the horse tries to duck away, he will be able to do so, but if the reins are left as described, the horseman has only to catch them together at the throat to have control.

The bridle is now transferred from the left hand to the right, still being held by the middle of the crownpiece. The horseman next puts the bridle in position on the head, the cheek pieces on either side of the jaws, the crownpiece and browband resting on the fore-head below the ears, the bit dangling against the teeth, it is here that the novice has his difficulties, for many horses refuse to open their mouths voluntarily to receive the bit.

Bridling the Horse

The left hand holds the bit in position while the right hand pulls the bridle into place after the horse has opened his mouth.

To overcome this the horseman must place his left thumb in the exact centre and at the bottom of the bit (if a snaffle is being used this is the point at which the joint comes) *the fingers of that hand reach under the chin and are inserted into the horse's mouth on the far side, coming to rest against the bars.* The taste of flesh is repugnant to the horse and will cause him to wrinkle up his lips and open his mouth. There is no danger of the horseman being bitten inasmuch as there are no teeth at this point.

As soon as the teeth are parted the horseman pulls up on the crownpiece with his right hand, at the same time guiding the bit into the mouth with his left. The crownpiece is then slipped back over the ears. The final step is to buckle the throatlash and fasten the curb chain. In fastening the latter be sure and untwist it until it lies flat, then with the thumb towards you, slip it over the hood. It should lie flat in the chin groove and the edge must not press

Placing the Bit

Fingers of the left hand enter the mouth at
the bars to cause the horse to open up and
allow the bit to be slipped into place.

against the chin. If you can slip three fingers under the chain and not disturb the position of the bit in so doing, your chain is loose enough for the average horse.

If the throat lash does not buckle readily it may be that the brow band has slipped down on the off side and must be slid back again. A good horseman always looks over his bridle and saddle from both sides before mounting to make sure that everything is adjusted

properly and to fasten any stray keepers, those little leather loops
that keep the ends of the straps from flying.

BRIDLING THE HEADSHY HORSE.

A few horses take special handling in bridling. This is almost
always due to cruel or inexperienced handling on the part of the
trainer. If the horse throws his head up, trying to avoid the bit, try
the following method: Hang the bridle over your left arm, take the
reins in two hands, six or eight inches from the ends on either side

Bridling the Headshy Horse

The hands are held down in order not to
frighten the animal. The next step is to pull
down, the crownpiece will act on the horse's
nose to lower the head when the bridle may
be slipped on in the orthodox manner.

and hold them in front of you to form a loop. Standing at the shoulder, hold this loop in front of the horse's head. Not seeing your
hands he will probably not resist, but will allow you to slip the reins
over his head. Next take the bridle in two hands in the same manner, holding the cheek pieces and with the crownpiece forming the

loop (see illustration, page 101). Slide this on in the same manner. If the horse throws his head before it is in place, by pulling the crownpiece downwards across the nose you can usually induce him to pull it down again. Horses with vices such as these should be bridled in their own stalls.

If you find it impossible to get the reins over the head in the manner described, unbuckle the snaffle reins, put them around the neck at the shoulder, rebuckle and slide them up to the poll from behind, this will give you some measure of control. Head-shy horses, if handled carefully and quietly by the same person, usually get over their fears as far as that one person is concerned, but often will not let a stranger handle them. It is a sad commentary on the trainer when a horse is so head-shy as to cause trouble in catching and bridling, it indicates rough treatment and that the colt was not properly gentled before being handled.

SADDLING.

Saddling is not as difficult as bridling, nevertheless incorrect saddling is responsible for about a third of the sore backs that one sees, bad riding and ill fitting tack being responsible for the other two thirds.

Having bridled the horse, the horseman prepares the saddle by seeing that the stirrup irons are run up on the leathers and that the girth is either entirely detached or else laid across the seat where it will not flop over and scare the animal. If a pad is being used this must be put on first. Lay it on the horse's back a little forward of where you want it to rest, now slide it back into position. This smooths down the hair. Next pick up the saddle, cantle in the right hand, pommel in the left. Place the saddle *very quietly* on the horse's back. If no pad has been used the saddle must be placed forward and slid back, but if the pad is already in place the saddle may be put on where it belongs. Some horses have the habit of wriggling the pad back as soon as it is put on the withers, in this case you will have to hold it in place while you put on the saddle.

When the saddle is properly placed, or around to the far side, pull the girth off the seat and buckle it to the billets. Remember that with a folded girth the open end goes to the rear, the fold to the front. It is less apt to rub the elbow if placed this way. Check under the skirts to be sure that nothing is out of place and that the pad lies smoothly. Now come around to the near side and buckle the girth. Don't try to get it as tight as it needs to be all at once, pull it up until it is snug and just before you mount, pull it up again. When the saddle is properly placed the girth will come about four inches back from the elbow. If the pad is the folded type it should extend a few inches in front of the pommel, if it is the shaped type it will have two straps which are to be slipped over the billets before the girth is attached. Leave the stirrups up on the leathers until you are ready to mount, and as soon as you have dismounted, slip them up again. It is both sloppy and dangerous to lead a horse with dangling stirrups which may fly out and catch on something, frightening the horse and perhaps causing a serious accident.

Cause and Control of Vices

CRIBBING.

This is a common vice and is incurable although it may be controlled. There is some disagreement as to the cause of cribbing and also as to exactly what the horse actually does while cribbing. Most horsemen are familiar with the cribber, he grasps the edge or end of a piece of wood or metal, he presses down on it with his upper teeth at the same time extending his head, tightening the muscle at the base of the throat and uttering a grunt. Horses have no vomiting muscles, and many authorities contend that in the act of cribbing the horse is trying to expel the gas in his stomach. They say that one cause of cribbing may be that of taking the foal away from its mother before it is ready to be weaned. Other authorities say that cribbing is a nervous habit, not a digestive one, that the horse is actually sucking in wind when he cribs. Certain it is that a horse will pick up the habit of cribbing when stabled next to a cribber, and once having picked it up, he will never get over it.

Although the vice cannot be cured it can be prevented in the following manner. Take a broad strap, one and a half to two inches wide and buckle it tightly around the horse's throat. This will prevent his extending the muscle which he uses in cribbing and so prevent the cribbing; it will not prevent his eating or drinking, as at this time he lowers his head and the strap is automatically loosened. A little experimenting will be necessary to get the exact degree of tightness; if too loose the horse will continue to crib, if too tight you will cut off his wind. Needless to say a horse should never be ridden in a cribbing strap.

TEARING THE STABLE BLANKET.

This is an aggravating vice which some horses seem to develop for no good reason, they reach around behind them, catch the blanket in their teeth and try to pull it off. There is a patented device known as a "cradle" which can be put on the horse to prevent this, but the following improvised arrangement is usually effective and all you need is an old rake handle. Measure the handle and cut it so that it will reach from the side ring in the halter to the surcingle. Now bore a hole one inch from each end, run short lengths of heavy cord through the holes and by these attach the pole at these two points. The horse can now raise and lower his head but he cannot bend his neck around to tear the blanket.

WEAVING.

This is a nervous habit sometimes encountered. The horse stands in his stall, head low, and shifts back and forth all day long. Keeping him in the paddock with other horses as much as possible, putting him in a box or a stall where he can look out helps. Such horses are usually hard to keep in any sort of good condition.

CROWDING.

Some horses have the bad habit of waiting until you enter the stall and then trying to crowd you against the wall. This is a vice easily cured by the following device. Cut a length of rake or broom handle a few inches longer than the breadth of your body. Slightly sharpen the two ends. Every time you enter the stall carry this in front of you, the two ends extending, as the horse crowds he comes up against the point and usually one or two lessons will be enough to teach him to mend his manners.

KICKING IN THE STALL.

High strung horses sometimes get the habit of cow kicking (kicking sideways) in the stall whenever they eat. Such horses should be kept in box stalls if at all possible, as they can bruise their legs badly. This habit can sometimes be controlled by the following method: Attach a small ball to a length of elastic or rubber cut from an inner tube. The elastic should be long enough to go around the horse's fetlock joint and allow the ball to dangle just clear of the ground. Each time the horse kicks the ball will fly out and bounce back on his leg with some force. This device is helpful in many cases. For the horse that has the habit of kicking at people or horses as they are led behind him, and on whom the above method does not work, I attach a broad piece of webbing across the back of the stall just above the hocks. This will not cure the vice, the horse will continue to kick, but it will prevent him from kicking high enough to do much damage.

BACKING OUT OF THE STALL SUDDENLY.

In handling horses that have acquired this vice you should always have a length of good strong chain across the back of the stall and the horse should also be fastened by the halter shank. When you go into the stall to take him out, refasten the chain behind you. Now undo the shank at the wall, not at the halter, so that you have the length of rope to hang onto. In all probability, as soon as the rope is undone, the horse will plunge backwards and come up hard against the chain. Speak to him sharply, lead him forward to his place and tie him up without taking him out of the stall at all. Go out of the stall and repeat the whole performance two or three times. When the horse stops trying to duck out you can undo the chain quietly and back him out a step at a time, leading him forward again if he tries to repeat his former habit. As a rule this method is very successful and only needs to be repeated a few times.

KICKING WHILE UNDER THE SADDLE.

A good horseman never hurts a horse *unintentionally*, but there are times when it is necessary to punish a horse and this is one of them. Before a horse kicks another horse he usually puts back his ears and tucks in his tail. When this happens get ready and the instant the horse kicks, pull his head up hard, at the same time giving him two or three good strong cuts with a flexible switch. You must be a good enough horseman to retain your seat if the horse attempts to buck, and switching him three seconds *after* the bad conduct is useless, punishment must come immediately. A great many horses will only kick with a beginner up and here you are defeated, for the beginner is not good enough to inflict the required discipline quickly enough, he is too occupied trying to stay on after the sudden motion of the kick; yet when the experienced horseman mounts, do what he will to provoke the animal into misbehaviour so that he can administer a lesson, his mount will wisely refuse to act in any except the most gentlemanly-like manner! Your only remedy is to mount beginners on such horses as seldom as possible and when you do have them ride a little behind.

KICKING WHILE BEING MOUNTED.

Many western-broke horses have this unpleasant habit. Never mount such a horse facing front. Stand well up at the shoulder, have the reins short and if the horse tries to cow-kick, pull up on them and slap him on the shoulder with a switch. When the horse stands quietly to be mounted, reward him with a bit of apple or carrot.

STARTING FORWARD BEFORE THE RIDER IS PROPERLY SETTLED IN THE SADDLE.

This is a common fault, entirely due to bad training and, fortunately, easy to cure. Take the horse to a quiet place where there is no temptation to get excited, an indoor riding hall is best, but any quiet corner will do. Arm yourself with two or three carrots or apples. Holding the reins short, mount your horse very slowly, checking any forward movement at once, hold him steady until you are comfortably settled, then reach forward and, from the saddle, give him a bit of carrot or apple. Dismount and repeat the performance over and over again. After the fourth or fifth repetition the horse will come to expect the tidbit and will wait for it voluntarily. The most restless horses can usually be broken in one morning by this simple method. Just be sure that your horse understands that he is not to make a move until *you* tell him to do so.

BITING AT ANOTHER HORSE.

This vice falls into the same category as kicking and requires the same immediate and prompt punishment. It is often more effective to switch the horse that bites on the side of the neck or across the ears rather than on the flanks, at the same time pulling him back and slightly away from the other animal. Many a horse will wait until the rider is relaxed and riding with a slack rein to try these tricks. Keep alert and have your horse in hand at all times and you should have no trouble.

CHARGING IN THE PASTURE.

This is a very disagreeable and dangerous vice. Ponies are especially apt to acquire it and it often comes from making too much of a pet out of the horse, particularly from feeding him out of your hand. The animal will start by nipping playfully and will get rougher and rougher until he ends by baring his teeth and running at you whenever you enter the pasture. Some will even wheel and kick as soon as they are within striking distance, or rear up and attack with the front legs. Such a horse should have as much work as possible and little grain. When you enter the paddock, carry a good stout stick with you and catch the horse across the nose as he comes at you. This takes nerve, but you have got to teach the animal that you are master and that he cannot bluff you. Many an animal has been speedily cured in this way, but, returned to the inexperienced or timid horseman, they sometimes revert to their old habits. Above all, never turn your back on an animal that has acquired this vice, you may be seriously hurt.

REARING UNDER THE SADDLE.

There are two classes of rearing horses, those that do it through vice and those that do it through fear. The latter should be left to the professional horse trainer. They must have a long course of re-education in competent hands; under heavy or inexperienced hands they may become so hysterical as to come over backwards with the rider. But the horse that rears just because he doesn't want to leave the stable or go down a certain road is bluffing. He will rarely rise more than a foot or so off the ground, often he will not try it at all under a strong rider, whereas, under the rider who is timid or uncertain, up he will go at once.

The rider must be capable enough to do two things, throw his weight forward and keep it there, at the same time applying legs or whip vigorously to urge the horse forward. He must have a short enough rein so that if the horse tries to go from rearing into kicking or bucking he can control him, but a pull on the mouth at the wrong moment may cause him to lose his balance.

Sometimes a horse may be broken of this habit by pulling the right rein under the right boot and stirrup and applying the left leg. This turns the horse in short circles to the right, the rein, being under the boot, gives a downward pull on the head which effectively keeps the horse from rising.

An old method used to be to break a paper bag of warm water between the horse's ears as he rises. The animal is supposed to think that the liquid is his own blood and be a reformed character from that day. One would have to know at exactly what moment the horse would give trouble and be prepared, carrying around a superfluous bag of water for hours might prove inconvenient.

BOLTING.

There is only one effective way to stop a runaway horse and that is to turn him in as short a circle as possible. Few horses are confirmed runaways, but an animal may get going a little fast in the hunt field or when galloping over the pastures in the early spring, and if the rider gets scared and loses his head, the horse does also. An alert rider can feel the horse getting overexcited and will check him then rather than wait until he is running. It is no harder to stay on a running horse than on a galloping horse, there is no need to get excited, especially if you have plenty of room. Settle yourself firmly in the saddle, take a good grip on the reins, but remember that the horse is stronger than you are, you cannot stop him by a steady pull. Bend his head to one side or the other, he cannot run forwards very fast if his head is turned to the side, and it is much harder for him to brace himself against the bit when the pull is from one side only. Many a horse that has the habit of pulling against the bit, not with the idea of running but just from freshness, may be controlled by alternately pulling and letting up on the reins.

If you are riding in company never, under any conditions, pursue the runaway horse of your companion. If you hear a horse coming up from behind, block the road by standing your horse across it, and try to grab the rein as he passes, but if the runaway is in front of you, *stop*. Nine times out of ten the bolter will also stop. If he doesn't, shout to the rider not to lose his head but to sit back and try and turn his horse. If the rider is thrown and the free horse goes on, turn your horse around and walk back a few steps. Usually the other horse, hearing your retreating footsteps, will turn and come back also. Dismount, have your companion hold your horse in place and try and work your way on foot so that the loose animal is between you and your mount. Talk to him in a quiet voice, pick up a handful of grass and coax him either to come up to you or to follow you back to your horse. An animal will sometimes come up more readily to a man on foot if the latter crouches or kneels so that he is lower than the horse and if he does not look directly at him. It is seldom that the above methods do not work, unless you are near home, in which case the miscreant will run back to the stable.

Two loose horses are another matter. They will egg each other on and run indefinitely. There is little you can do beyond explaining to the disgruntled riders who will now have to walk home that had they had the presence of mind to hold on to the reins when they fell this might have been avoided. If the horses do not return to the stable alone within a reasonable length of time (which they will very likely do) you will have to go searching for them with a pail of oats. Here is where your lessons of teaching your horses to allow you to come up to them readily will count.

SHYING.

Just as firm and sharp discipline is recommended for the kicker and the biter, so tact and gentleness must be used on the horse that shies from fear. Horses will sometimes develop a fear for some specific object usually traceable to incidents which have occurred during the period of training. Such a case is the one in which a three-year-old, out on the road under the saddle for the first time, was passing a standing truck filled with sand when the men, not seeing

im, threw out a shovelful in his face. The animal, now twelve years
old, still remembers the incident, and though he passes moving
trucks and other vehicles with no difficulty, is still hesitant about
going by one which is standing still.

Learn to know what scares your horse and then be on the lookout
so that you will not be unseated by his sudden movement. Talk to
him quietly, urge him forward firmly with your legs. Sometimes
allowing him to stop, smell and examine carefully the object of his
terror will give him confidence. With other animals it is best to turn
their heads away from the object which scares them and get by
quickly. If you are riding with a companion, let him go ahead of
you, your horse, especially if he is young, will probably follow
readily enough.

Many horses are afraid of stepping in water or mud or of cross-
ing bridges. Let them follow a companion or, if alone talk to them
and coax them on. Be prepared for a sudden plunge once the animal has gotten up enough courage to step on the strange footing.

REFUSING TO LEAVE HIS COMPANIONS.

Such a horse is said to be "herd bound" and the habit is very
common in colts. Other horses will employ this vice only when
mounted by timid or inexperienced riders showing that it is not a
question of fear but of stubbornness. The animal may rear or he
may simply plant his feet and refuse to budge. If he does the latter
the good horseman will turn him a little to one side and then to the
other until he has him moving, and then spur him on. "Rolling him
up," an old German cavalry method is often successful. Turn the
horse in short circles until he is slightly dizzy and not sure of directions, he will usually move off without further trouble. You may be
certain, however, that once having put up a defence of this sort,
the horse will wait until he thinks you are off guard and try it again.
Keep alert and urge him forward with a sharp word or a flick of
the whip before he has time to turn. If he does succeed in turning
in an attempt to rejoin his companions, don't try and pull him back
but keep him turning until he has completed the circle and is once
more headed in the desired direction. If this is done quickly and
the spur or heel immediately applied, the horse will often move
along willingly enough having become confused as to which way

he wants to go. Above all, never give up the fight or you will never master your horse and he will go from one vice to another.

Use these same methods in curing the "barn-rat," the horse who refuses to leave the stable yard except in company.

EATING GRASS UNDER THE SADDLE.

It would hardly seem necessary to mention this vice for which there is no excuse except that it is a very common one in ponies and also that many horseman who should know better, will often allow their pets to graze while they are sitting on them.

There are two reasons why this should not be allowed. In the first place, when your horse is working he should work, and if you let him play or eat he will very quickly try and see what else he can get away with. Secondly, when a horse eats, particularly when he eats grass, his belly expands making the girth too tight and often causing girth sores.

An adult rider should be strong enough and should stay sufficiently on the alert to keep his horse's head where it belongs, but many ponies have this habit and a child has not enough strength to do anything about it. There are two devices which may be successfully used to cure this habit. The first is a check rein. Put it on just as you would put on a check in harness, attaching it to the pommel of the saddle by means of a light rope or strap. This should run from one billet strap to the other or it may be fastened to the iron catches which hold the stirrup leathers. The objection to this method is that a clever pony will choose a bank on which the grass grows high and continue his meal.

Another, and perhaps more effective method is to procure a calf muzzle and put it on over the bit, allowing the reins to come out between the meshes. Run a rope or strap under the crownpiece and through the browband, attaching it on either side to the muzzle so that the latter is held firmly in position and cannot be rubbed off.

SHYING OUT AT JUMPS.

This is a defence habit rather than a vice and is usually the result of bad schooling. The horse is being asked to jump an obstacle too large or of a type to which he is not accustomed. However, in the hunt field, it is sometimes necessary to go over an unusual obstacle and to do it before hounds are so far away that you have lost them for the day.

Most horses have a tendency to shy to the left. A few will go to the right. In the former, instead of starting your horse for the jump directly opposite the centre of it, as one would normally do, start him from the left heading, on a diagonal to the right, almost as though you planned to jump the right post of the obstacle. Two strides from the jump, straighten the horse out so that he can take off squarely. In jumping a horse that shies to the right, reverse the procedure, coming in *from* the right. This is a handy thing to know as it nearly always works.

GETTING EXCITED AT JUMPS.

This habit again is the result of faulty schooling. There is nothing to be done except to reschool the horse for jumping as though he had never seen a jump before. Start with the rail on the ground and keep it under eighteen inches for at least a month. Have a variety of those low obstacles placed at random around the schooling ring. Change their positions daily and ride in and out among them as well as over them. Never allow the horse to know when he is to be asked to jump until he receives the signal to do so a stride from the jump. Keep the pace down to a trot at the most. Make a habit of riding in to an obstacle as though you planned to jump, and, at the last moment, instead of giving him the impulse, rein back, let him stand a moment, then turn and go off without taking the bar. Only increase the gait and the height of the obstacle when all signs of excitement have been forgotten. Reeducated in this manner horses formerly completely uncontrollable over jumps, may be so trained that they will even take to "mental hazard" jumping in which the jumps are such things as chairs, tables, rattling pails, tubs filled with water, etcetera.

REFUSING.

Again the result of faulty schooling. Reschool as suggested above. If your horse refuses in the Hunt field it is only ordinary etiquette to allow the rest of the riders to precede you, following at the end when your mount will, in all probability take the jump readily in order to keep up with his companions. If you are alone try backing a few steps and then starting for the bar at a distance of twelve or fifteen feet. Never give a horse that has refused a long run, it only gives him time to decide to stop again.

CHAPTER VIII

Riding

Riders may be classified as follows: Beginners, those who have done no riding at all and who both look and are completely uncomfortable in the saddle no matter how well mannered their mounts may be. The beginner is characterised by either a tendency to hunch over and cling like a monkey to the pommel, or to lean back, hands widespread and shoulder high in an effort to retain his balance by means of pulling on the reins. His legs are now here, now there, feet hanging any old way, knees agape. All riders have gone through this painful period; fortunately for the horse, under proper instruction it doesn't last long.

The intermediate class of rider may, and often does, look very well on his horse. He may hunt in a mild fashion or exhibit at horse-shows. He enjoys his riding and does no harm to his mount but, and here is the difference between this class and the third, the experts, the intermediate is at home only on a horse with whom he is familiar and which is perfectly mannered. Put him on an animal that comes out on a frosty day with a bit of stall courage and gives some light hearted bucks; put him on an animal with a sensitive mouth when he has been accustomed to a heavy mouthed brute, let him lose his stirrups while clutching for his hat and your seeming accomplished horseman grabs for leather like a beginner, all sense of ease in the saddle having vanished. Usually the horse is pretty sure of the lack of skill and uses such an emergency to advantage.

The great mass of riders belong to the above category. It takes many years of riding to become qualified for classification with the third and most exclusive group, the first-class horsemen. Many years of riding, much patience, great natural ability and a real love of horses, as well as a complete absence of fear where they are concerned.

But what a pleasure to watch such a horseman! He rides as though he and the horse were in very truth one creature, one unified mass of muscle under the control of one mind. Watch such a rider deep in conversation while trotting or cantering over the turf. His mount shies suddenly, the rider's body bends automatically with the motion, his hands steady the reins and return the horse to his place, all without interrupting his talk or seemingly paying any attention to his riding, as is indeed true, for years on the saddle have made his reactions purely automatic and subconscious.

Study this rider on a horse that has just been giving trouble under less experienced hands. With no apparent effort the erstwhile stubborn, rebellious beast goes through the most intricate movements as though reading the mind of his master. A moment ago, under the most vigorous kicking or slapping with the crop the same animal would hardly move faster than a snail's pace, each time he neared the exit to the ring he veered for it. He refused to stay out to the wall, but cut his corners shorter and shorter until he ended up in the centre and stubbornly stayed there. The poor rider, mounted for perhaps the third time, completely helpless.

See now, under the hands of the expert how completely differently he conducts himself! His whole bearing has changed. He no longer slops along, almost falling over his own feet, instead he holds himself with pride, well balanced from nose tip to tail tip. He positively hugs the wall, completely ignoring the exit gate. He takes an energetic trot or a gentle canter, executes figure eights, at the slightest hint from the master. In fact, to all outward appearances the latter is not doing nearly so much to make him behave as did the other rider who just dismounted. Unless you look extremely closely you cannot see that he is doing anything at all!

All of the above is merely a preliminary to ask you to take into your heart and engrave for ever on your mind the horseman's maxim, "It's never the horse, it's always the rider." To a beginner this may seem like a hard slogan. It is so much easier to lay the blame on the horse when he won't do as you want him to do, but think a minute and you will see that it is really a very encouraging motto, for all you have to do is to study and practice and learn, and eventually you will conquer the very things that seem so hard at first. Remember too, that no matter how many years you ride, there will always be something new to be learned. It is this which makes the art of equitation so fascinating.

ANALYSIS OF THE MODERN SEAT.

Our fathers sat back in the saddle, stirrups long, legs almost straight from the thigh down, hips well behind the feet. The weight was back on the centre or behind the centre of the saddle. When they jumped they carried their bodies perpendicular to the ground, sitting deep in the saddle from take off to landing, and simply extended the arms to give the horse more rein as he extended his neck. Study of ballistics has shown that such a distribution of weight places the greatest hardship on the horse and renders it more difficult for the rider to maintain his position when an unexpected movement occurs. It was discovered in racing, for example, that by riding with exaggeratedly short stirrups so that the jockey's weight was taken off the loins and put on the shoulders and withers of the horse the effect was as though the rider weighed eight or ten pounds less than he actually did. In a very short time all jockeys adopted this seat for racing although, exaggerated as it is, it is not practical for other types of riding.

Sit in a chair, your feet flat in front of you about eighteen inches apart. If some one jerks the chair out from under you you will inevitably land on the floor! If you sit this way on a horse (see page 21) and your horse moves his hindquarters up, down or sideways suddenly, you will find yourself on the ground. Now stand on the floor, your feet apart and toes slanted slightly out, about the way they would be for walking, bend your knees so that your body is lowered about four inches, at the same time bend your ankles so that you are standing on the inside edges of the soles of your shoes. If it were possible to put a horse between your thighs at this point so that you found yourself in the saddle without changing your position beyond the fact that you would want to drop your heels a couple of inches, you would be sitting correctly. Your ear, your hip bone, your ankle bone would be in a line. Your stirrup leather would hang perfectly straight. Your knee would "cover" your toe so that when you looked down you could not see the latter. The inside edges of the sole of your boot would be pressed against the

Typical Beginners' Seat

Hands are too high and flat, legs too far forward, weight too far back.

Improved Position

This is a little better but the weight is still too far back and the stirrup leathers do not hang straight, showing the feet to be too far forward.

Good Position in the Saddle

Rider's ear, hip and ankle are on a line and hands are held correctly with a straight line from elbow to bit.

inner side of the iron and a person standing on the ground would see that the tread of the iron did not hang flat but was slanted so that the outer edge was higher than the inner. Your body would be erect but not stiff. There would be from four to six inches between the end of your buttocks and the cantle of the saddle. You would not be resting on the back of your spine, but on the ends of the pelvic bones, the so-called "sitting bones." Your thighs and your lower legs from about an inch below the boot tops to the knee would be in close contact with the saddle but there would be no forced pressure. The term "grip with the knees" is very misleading. What it means is that the rider maintains his seat with his legs, not by use of his hands, but the tyro thinks that it means that he must grip the saddle with all his might at all times, such a practice tightens the thigh muscle forcing the rider out of the saddle and prevents the relaxation that is essential in any form of activity. It is true that at times there is increased pressure of the thighs, as the horse takes off for the jump, for example, when the rider is intentionally out of the saddle, but it is never continuous. Normally the rider maintains his seat by balance, not grip. Having assumed the position above take up the reins holding them in both hands. The knuckles should be vertical, the reins should enter the hands between the fingers and go out under the thumb, the latter being on top. Imagine that you are holding lighted candles and the bight where it is covered by the thumb is the flame, and you will have your hands in the correct position. The distance between the two hands varies with the thickness of your horse's neck and shoulders, but they should always be held several inches *in front of* the pommel of the saddle, never over or, worst of all, behind it, and about three inches above the withers. A straight line should run from the point of your elbow, down your arm and the reins to the horse's bit. The reins should not touch the horse's neck, so you will have to keep your hands from six to ten inches apart. If you are using a pelham bridle or a full bridle your snaffle reins will enter between your little finger and your ring finger, your curb reins between the ring finger and the longest finger. The light pressure of your fingers on the *edges* of the reins will keep them in position, plus the pressure of your thumb on the bight. Your fingers, wrists and arm should be relaxed and flexible, ready to give or take as the movements of the horse demand. If you have not had much riding ex perience you will find it extremely difficult to maintain a light contact with your horse's mouth at all times, not allowing your reins to become too loose, nor pulling too hard to maintain your balance. Your ability to accomplish this skill will depend entirely on the security of your seat.

This is bad position of both hands and legs. Notice the daylight between the rider's knees and the saddle.

Here is a good position of legs. (Compare with photo at left.) The horse seems happier too.

To test the correctness of your posture, drop your reins for a moment and, folding your arms across your chest or putting them on your hips, stand up in the stirrups keeping your knees still a little bent and bringing the crotch almost over the pommel. If you have taken the right position, well forward, you should be able to maintain this position easily not only while your horse is standing, but also while he is walking, trotting and cantering. To practise taking and maintaining this position without the use of your hands to balance, for several moments at a time and at the different gaits is one of the best exercises that can be found for the purpose of developing balance and co-ordination. Be careful that you do not allow your body or legs to become stiff, nor your heels to fly up. If you find it hard it is because you are not properly balanced. At the trot the rider leans a little forward and rises off the saddle at each alternate step. This takes away the jolt of the gait, the joints of the ankle, knee and hip acting as shock absorbers. Be sure that the action of posting is done by the leg muscles alone, don't allow the back muscles to enter into the picture, they should remain relaxed. When the rider rises as the horse moves his right foreleg and left hind leg forward he is said to be posting on the "right diagonal," by standing in the stirrups for a beat, or bouncing once in the saddle he changes diagonals and posts on the left. Most riders have the habit of using either one or the other diagonal continuously, it is wise to switch over occasionally to rest the horse and to develop his muscles. In riding in a small ring it is best to post on the outside diagonal, thus the back leg of the horse that is on the inside takes up the weight and acts as a support to maintain the balance on the corners. Many riders "get behind the trot," instead of anticipating the movement and rising with it, keeping well forward in the saddle, they wait until they are propelled out of the seat by the action of the horse. This is wrong and puts an unnecessary strain on the animal.

The most common fault at the trot is allowing the hands to bob up and down, thus alternately loosening and tightening the reins. A horse holds his head very steady while trotting, he does not bob it the way he does at the walk. He holds it somewhat higher so it is necessary to shorten the rein slightly, the rider, in the act of posting, raises his hands, to avoid this he must slightly straighten the elbow joint at each step so that his hands may remain in the same relative position to the horse and not follow the movement of the body. An excellent exercise for this is to extend the little fingers and touch them lightly to the horse's withers, one on each side about four inches below the top, now maintain that light contact, neither pushing down on them nor allowing them to leave the horse while you trot for ten minutes or so. It will make you conscious of the necessity for a flexible elbow and steady your hands very quickly.

At the canter or slow gallop the rider keeps down in the saddle but not too far back. Here the motion is backward and forward rather than up and down and the muscles of the rider's back come into play bending slightly with the movement so that the rider has the feel of "polishing the seat of his pants on the saddle." The tendency of the inexperienced rider is to tighten his thigh muscles, draw up his legs in an attempt to clutch the horse's sides or else put too much weight in the stirrups, either procedure causing him to bounce several inches at each stride. This is extremely uncomfortable for the rider and also for the horse.

If you have a horse that can be depended upon to take and keep a gentle canter without too much help from you, and one which does not become excited by feeling a shifting of weight and a tendency to cling with the calves, the best way to acquire a relaxed seat at the canter is as follows: Put your horse into the gait, as soon as he has steadied down, take your feet out of the stirrups letting them hang their full length and as completely relaxed as possible. You may find it necessary at first to steady yourself with a hand on the pommel. If so grasp the latter with the hand that is on the *inside* of the ring (the right hand if you are going to the right) keeping the reins in the other hand to prevent the horse from cutting in. Be sure that the hand holding the reins is not drawn up but is held well down and a little to the side, only enough pressure being put on the horse's mouth to insure his control. Keep your weight a little back, not too much, and consciously try and let your body swing with the motion of the horse. Try and catch your stirrups while still cantering as the break from the canter to the trot is rough. If this is impossible, pull your horse in to a walk, rest a moment and then repeat the whole thing. Gradually you will attain sufficient balance so that you can let go the saddle entirely and can catch or drop your stirrups easily at any point. When this point has been reached, tie your reins in a knot and as soon as the horse has steadied, drop them as well as the stirrups, reaching down for them only if the animal goes too fast or cuts in. Keep your hands on your hips or fold them. When you are comfortable without your stirrups try going without the saddle entirely. You may want to buckle a stirrup leather around the neck of your horse at the withers if he has no mane on which you can tug should you lose your balance unexpectedly. Practicing in this fashion without reins or stirrups will do much to supple your back muscles and to give you confidence.

At the full gallop the rider takes the weight entirely out of the saddle, putting it on his knees and to some extent on his stirrup irons. He pushes his knees well forward and down, they and his ankle and hip joint take up the full motion of the horse, shoulders, head and back remaining steady and well forward. Make a practice of keeping the eyes up so that the head does not drop down. The reins must be shortened and the rider's hands are extended forward and well down on either side of the neck but without touching it. This is also the position that is assumed for jumping, by being completely out of the saddle the rider is not thrown off balance by the sudden motion of the horse, at the same time the horse has the utmost freedom to maintain his balance, and propel himself forward or over the obstacle. The rider's hands must be light on the reins, adjusting themselves to the movements of the horse's head, steadying, controlling, maintaining contact but never jerking or pulling at the wrong moment. Each hand works individually so that the rider can give or take on either side independently of the other. The objection to the "bridged rein" where the bight from each hand is held in the other is that it does not allow this free play and consequently restricts the movement of the horse's head. In jumping many army men like to take the reins into the hands in the following manner, the rein is grasped so that it enters the hand between the thumb and forefinger and the bight comes out under the little finger. Thus there is less chance of jerking the animal's mouth if the rider gets "left behind." A common fault at the gallop is to allow the legs to go too far back and to maintain balance by a steady pull on the horse's mouth. The rider should not attempt the gallop until he has become thoroughly experienced at the other gaits and then he should first practise taking the correct position when the horse is at a slow canter.

COMMON FAULTS IN RIDING.

THE HEAD.

    a. Carried too far back, chin extended.
    b. Carried low, chin sunk, eyes fixed on ground.
    c. Carried to one side.

THE SHOULDERS.
- a. Carried high in a rigid, strained fashion.
- b. Allowed to move up and down at the trot.
- c. Slouched over.
- d. One higher than the other.
- e. One more forward than the other throwing the rider off centre.

THE BACK.
- a. Humped over or held stiffly.
- b. Back muscles used to aid leg muscles in posting.
- c. Back muscles not used to take up motion of canter.

THE BUTTOCKS.
- a. Tucked under so that the rider sits on his tail bone instead of on the pelvic bones.
- b. Pushed back against the cantle of the saddle.

THE THIGHS.
- *a. Not flattened against the saddle.
- b. Turned incorrectly so that the rider's knees turn out.
- c. Muscles not relaxed resulting in a "forced" seat.
- d. Not keeping constant and even contact with saddle.

THE KNEES.
- a. Turned out so that the back of the knee and calf come in contact with the saddle rather than the back.
- b. Held loosely leaving a triangle of daylight between knee and saddle, a universal fault of beginners.
- †c. Not pushed ahead of the stirrup leather.

THE LOWER LEG.
- a. Pushed too far forward or too far back.
- b. Not flexible. The rider *must* be able to move his lower legs independently of each other at all times.

---

* It takes some time to flatten the muscles of the thigh so that they adhere closely to the saddle. Until these muscles have lost their natural roundness, the rider will be insecure and uncomfortable in the saddle.

† This is caused either by too long stirrups or by the rider having slid back in his seat. Stirrups should be adjusted so that when the rider takes his feet out and lets them hang, the tread of the iron dangles against the ankle-bone. They can be an inch shorter than this for jumping

ANKLES.

   a. Bent out instead of in.

   b. Too far forward or too far back.

HEELS.

   a. Level with or higher than toes.

   b. Pushed into the horse's sides. (A good horseman never touches his horse with his heels except for a purpose.)

FEET.

   *a. Pushed home in the stirrups, i.e., all the way through so that the tread comes under the instep.

   b. Thrust too far forward.

   c. Allowed to swing back and forth with the motion of the horse.

ELBOWS.

   a. Carried out and allowed to flop. To cure this, practise carrying a bundle of straw under the upper arm, especially at the canter.

WRISTS.

   a. Held stiffly.

   b. Held with the broad part horizontal to the ground instead of perpendicular.

   c. Bent too far in or out.

FINGERS.

   a. Inflexible. It is through the fingers that the rider feels the re-actions of the horse to the bit and consequently is able to judge what to expect. The acts of starting, stopping, turning, and changing gaits are done by slight movements of the fingers and wrist rather than the arm. They *must* be flexible.

HANDS.

   a. Held too high or too low.

   b. Held too close together or too far apart.

   c. Allowed to bob up and down.

   d. Held too far back so that the rider has "nowhere to go"; when he wants to pull in or rein back.

* Pushing the foot home takes away from the flexibility of the ankle joint and foot as well as constituting a serious danger. In the event of a fall the rider with his feet home runs the risk of being dragged as his toe may catch in the top of the stirrup iron.

Keep in mind that the horse is trained by the "reward and punishment" system. Thus, when you pull on the bit to tell him to slow down you make him slightly uncomfortable. By relaxing your reins as soon as he has complied, you reward him. The mediums through which a rider commands his horse are known as:

THE AIDS.

These are the hands (reins), legs, weight and voice when used to control the horse. The aids act in combination, each being ready to correct the result of the use of the other. Thus, if a rider urges his horse forward with his legs, his reins act as a control to prevent the motion from being too sudden or too fast.

TO START. Pick up the reins so that you have light contact with your horse's mouth. Press both legs slightly. It is permissible to speak to your horse but not to "click" to him, especially in company as this will affect the other horses as well. The weight is carried slightly forward. All these things tend to make the horse move off in a quiet, collected manner without undue excitement. They tell him that you know your business and will take no nonsense.

IN STOPPING. Use the fingers alone, the reins are tightened slightly, the rider sits down in the saddle and throws his weight slightly back, he may or may not speak to his horse. The object is to decrease the gait slowly so that the horse comes to a balanced halt and is not thrown back on his hindquarters. He should finish with his feet under him and his head in a normal position, the reins are then relaxed as well as the rider's legs.

TO TURN TO THE RIGHT. Press the left leg lightly against the horse's side, lean slightly to the right, carry both hands to the right so that the right rein acts as a "leading" rein and the left rein acts against the left side of the neck as a "bearing" rein. By using the leg and weight as well as the reins to cause the horse to turn you will make him turn his whole body in unison, his hind legs following the track of his fore legs. If you use only the rein, a common practice, he may only turn his head, swinging his hindquarters out and off the track.

TO CAUSE THE HORSE TO MOVE HIS HINDQUARTERS. To the left, employ the right leg a little back of its normal position, use the right rein as the rein of opposition (to the left with slight tension to the rear), have the other leg and rein ready to control the movement.

TO BACK. Collect the horse with the reins so that the head is bent at the crest and poll, the chin in. Press slightly with the legs to induce movement at the same time give light, vibrating pulls on both reins to cause him to move backward and not forward. The weight should be back. The horse should not rush backwards, but should move one step at a time as the rider demands, the reins being relaxed after each step as a reward. The movement should be light and active, not heavy, awkward or unwillingly. The horse should always be moved forward a few steps after being backed.

TO TAKE UP THE TROT FROM THE WALK. The rider shortens the reins a few inches to allow for the raising of the horse's head and squeezes with the knees. The degree of pressure needed will depend on the sensitiveness and training of the animal. The weight is carried forward and the rider rises up and down in cadence with the pace.

THE LEADS.

A horse is said to be cantering to the left when the left front and left back feet touch the ground in advance of the right. It is important that the horse be on the correct lead when turning sharp corners. If he should turn abruptly to the right while traveling on the left lead there would be no back leg under him to support the weight and he might fall. A horse that takes up the canter with one lead in front and another behind is said to be "disunited." This is uncomfortable for the rider and bad for the horse.

TO TAKE UP THE CANTER OR GALLOP FROM THE WALK, LEFT LEAD. The horse's head is turned slightly to the right, putting his left foreleg ahead of the right and "opening" the left shoulder. This is done by using the reins and left leg as directed in "turning," see page 26. The rider then urges the horse into the canter with the *right* leg only, at the same time giving the head a slight "lift" with the reins and then relaxing them. The weight is well forward. If legs, weight and hands work smoothly and in unison the horse will take the lead without excitement and without the insertion of any steps at the trot. As soon as he is in the canter he should be straightened out so that he is not looking away from the direction in which he is cantering. Some horses are trained to take the canter on the leading rein instead of the bearing rein, the head being turned in towards the centre of the circle. But the method I have described is more common in this country and is easier to teach inasmuch as the very position of the animal almost automatically insures the correct lead. TO DECREASE THE GAIT. The rider's weight and reins are carried to the rear. If decreasing from a trot to a walk the rider ceases to post and sits the last few trotting steps. As in the halt the decrease of gait should not be sudden or jerky.

THE JUMP. The rider assumes the position described on page 123, the FULL GALLOP. As the horse approaches the jump the pressure of the knees and thighs is increased and the weight carried more forward. At the exact moment when the rider wants the horse to take off he squeezes suddenly with knees and calves or applies the heel lightly, at the same time extending the arms well forward in order to give the horse all the rein he wants. The hands should not be extended to such an extent that the rider loses all contact with the animal's mouth, this is an invitation to the horse to stop or attempt to shy out. The moment for the take off signal may be easily judged if the rider will make a practice of counting the strides of his horse as he approaches the jump thus: one..two.. three..four..off, the "off" not coming at the same time as the "four" but following it in rhythm. As the horse lands the rider still stays out of the saddle in order not to put any additional weight on the hindquarters, only coming back when one or more strides have been taken. If there is any feeling of "jerk" the rider is behind the horse at the jump and is most probably damaging his mount's mouth. For beginners it is a wise precaution to put a stirrup leather around the animal's neck and let them hold on to this until the motion becomes easy. Keep the jumps low at first, under two feet, and only increase the height when you are sure you are in unison with your horse.

EXERCISES TO INDUCE CONFIDENCE AND RELAXATION.

The beginner, or the person who has not ridden in some years is apt to be stiff. His muscles need developing as does his courage. His fear is not so much that of falling as of what the horse may take it into his head to do, and of whether he will be able to cope with the emergency. The following exercises are based on those used by the United States Cavalry and are most useful in attaining relaxation, flexibility and confidence as well as in developing the riding muscles. They should be practiced constantly.

Exercise 1—The rider drops his reins and lies back on the horse, his head on the animal's rump. He should remain there for a moment, completely relaxed with his hands folded across his chest. At the command "forward," given by the instructor, he raises himself without the use of his hands and leans forward until his forehead touches the animal's mane. Repeat ten times. This exercise develops all the muscles, especially those of the abdomen, supples the back and builds confidence. Start with the horse at a standstill, later try it at a walk and trot.

Exercise 2—The horse being stationary the rider takes his feet out of the stirrups, swings one leg across the saddle in front of him, and, continuing to turn, swings the other leg across the animal's rump so that he is facing backwards, he then completes the circle, finishing by swinging the last leg across the pommel which puts him back in his original position. This exercise is for balance and to promote confidence, teaching the pupil that he can get back into position even though he may be out of the saddle and off balance. The beginner will have to use his hands in turning, later he can learn to spin around with his arms folded and also to do the exercise on a moving horse.

Exercise 3—Putting both hands on the pommel and dropping reins and stirrups, the rider puts his weight on his hands and brings his feet up on the seat of the saddle, he then stands upright. If this is found to be too difficult at first he can begin by standing in one stirrup, putting the other foot on the seat while holding to the pommel and cantle with both hands. He then brings the other foot up, making sure to place both so that they are pointed forwards, and stands up. This exercise is primarily to promote confidence.

Exercise 4, rotating arms, head and legs—These movements are to develop flexibility and relaxation. They may be used in any order, a suggested sequence is as follows:

Reins being in the right hand, left hand on the waist, horse moving in a big circle to the left, the rider executes the following movements, extends the left hand in front of him and looks at it, over his head and looks at it, to the rear and looks at it, down and touches his left toe. The movements are repeated several times, varied by having the rider touch the right toe instead of the left.

The direction is then changed (change of hands) by riding on a diagonal across the ring and taking up the movement in the op-

Standing on the Saddle

This instills confidence and promotes a sense of balance. Notice that although the pony is kicking at a fly the young rider is still relaxed and balanced.

posite direction, reins are shifted to the left hand and the right arm is rotated as described.

Next drop stirrups, look up at the ceiling, look down at the ground, look over the right shoulder, look over the left shoulder, look up at the ceiling, at the same time catch the stirrups. Repeat from beginning. This exercise relaxes the neck and back and teaches the rider to catch the stirrups without the use of his eyes or hands.

Having again dropped the stirrups, stretch the left foot as far forward as you can, swinging from the hip, now swing it back, repeat five times. Do the same with the right foot. Next work both legs at the same time, one going forward as the other comes back in a scissor movement. Hold one leg away from the saddle and rotate the toe. Next rotate the whole leg from the thigh, repeat with other leg. Do both together. Swing both legs back, bring them forward and in front of you, clapping your heels together over the withers. These exercises promote balance, confidence and control of the legs individually.

Exercise 5. Learning to fall off—The beginner's greatest fear is that of falling, yet it is easy to learn to fall properly. A rider is seldom hurt if he is thrown *suddenly* from a horse as he usually lands on his back and instinctively curls up and rolls, he does not have time to stiffen or put out a hand to save himself. The time when a rider is most apt to get hurt in falling from a horse is not when he is *thrown,* but when he loses his balance and falls slowly, giving him time to put out an arm in attempt to save himself. This often results in a broken wrist, upper arm or collar bone. Of course riders are sometimes hurt by having a horse fall with and on top of them. This rarely happens unless the rider has the misfortune to have his feet caught in the stirrups and so is not thrown clear, another danger in falling is that of being thrown against a wall or fence. Actually the distance one falls off a horse is very little, three feet at most if measured from the soles of the feet to the ground. There is nothing to fear in falling this slight distance if the rider will teach himself to come down feet first. At the same time, he should be prepared to move forward on landing if the horse is still in motion.

The horse, being at a walk, the rider should drop his stirrups and take his reins in the left hand. He should then vault off landing facing to the front, not to the side, and holding on to mane, neck or saddle for balance. He should hit the ground on the balls of both feet, the knees being slightly bent to form a spring and absorb the shock. Practise this vault first while your horse is walking, later at the trot and the canter. Continue to do it daily until you automatically throw yourself off your horse should an emergency arise, holding the reins so that your horse does not go home without you.

RIDING WITHOUT STIRRUPS.

The seat at the canter without stirrups has already been described, see page 122. Bareback riding at all gaits is the quickest method of giving the intermediate rider balance and a close seat. This is borne out by the fact that all armies make their recruits ride bareback and without stirrups at frequent intervals during their training period. If indulged in too much the rider may become sloppy as to form, so one needs work with stirrups as well. In this riding one should be particularly careful to keep the whole body completely relaxed, failure to do this will prevent the rider from staying close to his horse at the trot and the canter. The rider should neither hunch over nor draw his legs up, he should sit directly behind the withers and not on the loins. He should practice all the gaits as well as low jumps and if possible should not use his reins but ride with his arms folded. It is very important that he retain his seat by balance alone, not by gripping with the legs nor tugging on the reins. See the illustration on page 8 of the young rider on the English Shetland.

DO'S AND DON'TS TO SAVE YOUR HORSE.

Don't bring your horse out of the stall, mount and trot or gallop off *á la* cowboy picture. *Do* walk him quietly for at least ten minutes so that blood can start circulating slowly. By the same token *don't* trot or canter at the end of the ride, walking the first mile out and the last mile in is a good old English maxim to which one should strictly adhere.

Don't take the saddle off a hot horse as soon as you dismount. *Do* loosen the girth and then walk him leaving the saddle in place, the same applies to the driving collar and saddle.

Don't jerk your horse's head up if he stumbles, he needs complete freedom so that he can recover his balance, *do* keep a steady light hold on his mouth at such times.

Don't throw your weight backward going either up or down hill. *Do* keep well forward, the steeper the hill, the more forward you should be. This leaves the horse's hindquarters free and helps him to dig in his toes.

Don't trot or gallop your horse on very steep, very rocky or very hard surfaces. *Do* allow him to pick his way down such places but with a firm hold to check any desire he may have to rush the last few steps. Harness horses that are used much on hard roads are usually shod with rubber. Furthermore they do not bear the weight of a rider, and so may be permitted to trot on the ordinary hard surfaced road.

Don't be irregular as to schedule or amounts in feeding your horse. *Do* arrange matters so that he may be fed a regular ration promptly, no horse will do well if fed at four o'clock one day, seven the next and five-thirty the following.

RIDING ETIQUETTE.

In riding alone be sure that you do not trespass on other people's property without their permission. Be careful not to step on lawns or to ride across planted fields. Close all gates firmly behind you. If you have the misfortune to take off a rail or knock down a wall, either repair the damage on the spot or report to the owner and offer to pay for having it repaired. In riding through traffic obey the signals and, as far as possible, indicate your own intentions as to which direction you are taking and don't hold up faster moving vehicles any more than you can help.

In riding with a companion the rider with the horse having the fastest gait should adjust his pace to suit his friend. There is nothing more annoying than to be riding a horse with rather a slow walk and have to jog continually to keep up with another's faster-gaited animal. In passing through brush and under low limbs do not attemp to hold the branch back, there is at least six feet between you and your companion, i.e., the croup of your horse and the forehand of his, trying to hold back a limb will only result in having it swing in his or his horse's face. Instead, brush through, letting go the branch as soon as possible so that it will be in its normal position when he gets to it. If you want to hold back a gate for some one else go through it first yourself, then turn and ride around to the other side of it, facing your companion.

Don't "click" to your horse or wave a stick, your companion's horse may think the signal or punishment intended for himself and act accordingly.

If you wish to pass another rider, don't come up from behind at a gallop or a fast trot. Match your gait with his, ask permission to pass, and wait until he has pulled aside before you attempt to do so. Ignoring this rule may cause you to get kicked or the horse of the other person to run away. If some one wishes to pass you, *back* your horse off the trail so that his head is towards the rider passing. This applies also if a motor vehicle passes you on a narrow roadway.

If your companion has to dismount for any reason, wait until he is mounted before going on, it is difficult to hold a horse still if another horse is in the act of leaving.

In riding with a group, particularly if you are in the lead, remember that the other horses will tend to go at the same rate of speed that you do, therefore adjust your speed to the ability of the poorest rider, looking back frequently to see that every thing is all right. This applies particularly when going down steep hills or over rough ground. A horse takes shorter steps at such times, if you allow your horse to take his normal walking stride as soon as he reaches the bottom of an incline, by the time the last few riders are halfway down they will have lost distance and the horses may break into a trot to catch up. It goes without saying that one should never *increase* the gait at the foot of a hill until the last person has reached level ground. In going through a gate or bar-way where you are leading and the last rider is to close the gate or put up the bar, halt the group as soon as all are through and keep them halted until the job is finished. You will find it easier to hold your horse steady if you turn and face the other riders, they, too, will be less anxious to go. If it is necessary for a poor rider to dismount and you wish the group to continue ahead at a walk it is better if a good rider remain with him to hold his horse while he mounts. In catching up the mediocre rider should keep *behind* the better rider and the latter should keep his horse's pace down to a slow trot. Too many times this rule is ignored with the result that the horse of the beginner comes up on a canter crashing into the end of the column ahead and causing all their horses to get excited.

ETIQUETTE IN THE HUNT FIELD.

There are many and varied rules pertaining to hunting. The following are a few of the most common and should always be strictly adhered to:

On arrival it is only polite to introduce yourself to the Master of Hounds, thanking him for inviting you to hunt. A man invariably tips his hat to the Master. Also you should not fail to thank him at the end of the ride.

When hounds have been cast and are busily working out a scent, the Whips posted and the field listening with all ears for the first signal of a find, there must be no talking or whispering and no unnecessary movement. Keep your horse and yourself as still as possible.

The worst crime a person can commit is to step on or otherwise injure a hound. Never take a jump if there is the slightest chance of there being a hound on the far side. If a hound comes up from behind, pull out of his way and call "'Ware hounds" so that the rider in front will be warned and can do likewise. If your horse has the unfortunate habit of kicking at other horses or hounds you must do more than tie a red ribbon on his tail, you must never allow him to get into a position whereby he might do damage. This means watching the horses of the other riders as well as your own.

The Hunt servants, the Master and the Field Master always have the right of way. They ride ahead of the field, behind or to one side of the hounds, you must not on any account pass them. If the wrong cast has been made and it is necessary to reverse the field, you will hear the cry "'Ware Whips" or "'Ware Master" and should back your horse out of the way, keeping his head towards the ones passing you. The Field Master is in charge of the riders, take any suggestion he may make as law.

In hunting country where the panels are narrow great care must be taken to avoid accidents. Never set your horse at a jump until the rider before you has cleared the jump and gone on, keep your horse *standing* a reasonable distance away from the panel until the other rider is well out of the way, you will find this easier to do if you turn your horse sideways, his head turned slightly away so that he does not see the rider ahead. Often two riders approach a jump perhaps twenty feet apart, a safe distance one would think, and then the leader slows up, stumbles or pecks and the second rider is practically on top of him if not completely so.

If your horse refuses at a jump pull entirely out of the way and don't put him at it again until all the others immediately following you are through. It is permissible to pass other riders while galloping on (Hunt Servants, etc. excepted) but do so only where you have plenty of room.

If a rider has been thrown and you hear or see a free horse coming up from behind, do your best to catch him as he passes, for such a horse can cause serious accidents. In going through gates it is not necessary for you to hold it open for the field, but neither should you allow it to swing back in another rider's face.

COMFORTABLE CLOTHES.

The kind of clothes you wear while riding will depend largely on what kind of riding you are doing. The loose blue jeans, cotton shirt and bandana are fine for western riding, worn for riding in the east where the rider posts it will be found that the legs of the trousers work up and are very uncomfortable. Some people prefer breeches and boots to jodhpurs, although many others like the latter. Whichever you choose be sure that there is room in the seat but that the crotch is not so long that the breeches form a fold across the saddle in front of you and so prevent you getting down where you belong. They should be as tight at the knees as possible and still allow you to bend the knee sufficiently to mount. Breeches that are too short in the inside seam tend to ride up out of the boots, this may be remedied by the use of "extensions" additional lengths of material not necessarily exactly matching, which are added at the bottom. The coat or jacket should be loose enough in the shoulders to permit free play, and *must* be split at the tails so that you do not sit on it, most riding coats are lined with rubber at the tails to protect them from the sweat. Sleeveless jackets, bright red or green or yellow washable jackets are all bad taste as are the fancy coloured jockey caps one sees, better wear no coat at all for informal riding in summer than these. Men's tweed sports jackets are excellent for country riding provided the tail is split. Whipcord, gabardine or cavalry twill in various weights depending upon the climate, are used most often for the breeches and jodhpurs, with various shades of tan, fawn and grey being the most popular shades. Professionals wear a special cut of black or navy jodhpurs and coats for showing in the saddle classes as do many riders in the horsemanship classes at night. Flannel or cotton shirts with either a regular, four-in-hand tie or a silk or wool coloured stock are best for country riding. For formal shows or in the hunt field one wears a white stock. It is interesting to note that stocks were first introduced as an article of clothing which might also be used for a bandage in an emergency.

The fit of one's boots is most important. They should fit snugly in the foot but loose in the heel. When new the tops should come just below the bend of the knee at the back, they will drop another inch or two as they wrinkle from wear. They should be as snug as possible at the calf, but if they are to be worn both in winter and summer they should be tried on over the heavier breeches to make certain that they will not be so tight as to cut off the circulation. Boots to be worn in winter should be plenty big enough in the foot so that several pairs of socks may be worn. People with very high insteps may find it impossible to buy boots which they can get on and off without a struggle. Field boots, those with laces at the insteps are a solution for this type of foot although they are not considered correct for really formal riding. Brown boots are considered correct to wear with a tweed coat, black with a melton. Gentlemen wear black boots with lighter tops for hunting, women's hunting boots have patent-leather tops.

All recognised hunts have their own hunt uniforms. If the conventional "Pink" coat is worn by the men it will have a collar of a special colour which is the colour of that hunt. Sometimes the men's coats are green or blue. White buckskin breeches or very light fawn are worn and a silk hat or a velvet hunting cap. The vest is usually yellow or plaid. Women wear oxford grey, navy or black coats with the hunt collar and a silk hat or bowler. A woman M.F.H. (Master of Hounds) may wear the coloured coat and the velvet hunting cap if she wishes. Both men and women wear the white stocks and leather gloves, carrying a pair of string gloves under the billet straps for wet weather. Their coats have the buttons of the hunt. A new member of a hunt may not wear the hunt uniform until asked to do so by the Master. Likewise if you are invited to hunt with a strange hunt you do not wear your own hunt uniform unless asked to do so. Instead you wear an oxford coat, fawn breeches and a bowler; this applies to both men and women.

During the cubbing season which precedes the regular hunting season, the members of the hunt wear "rat-catcher" clothes, ordinary tweed coats, coloured stocks, bowlers, etc. It is at this time that litters of young foxes are broken up and encouraged to run, and that young horses and hounds as well as young riders get their first taste of the grand sport of hunting the fox.

CHAPTER IX

Driving

The harness is put on the horse in the following order: The collar,* the hames, the bridle, the pad and breeching (be sure to run the girth through the breast plate or martingale before fastening the latter).

The traces, holdbacks and reins are left tied up until the animal is between the shafts. Failure to do this has caused many a runaway through the horse becoming frightened by the flying ends of the straps. After being backed between the shafts the traces are fastened first, then the holdbacks followed bellyband and check rein. The reins are now unwrapped and run back through the terrets. If the horse is to stand for some time the check may be left unfastened and the reins folded at the stitching and pushed a few inches through the off terret on the pad.

The bellyband should be more snug in a two-wheeled vehicle than in a four as its purpose is to prevent the shafts from rising. In double harness it is left quite loose.

The coupling reins in the double harness must be adjusted in such a fashion that the horses are not too far apart, neither do they travel with their heads towards each other. If, from bad coupling, horses have gotten in the habit of holding their heads together, a simple remedy is to reverse their positions.

The outside traces of a pair are fastened before the inside traces, this is to prevent the animals from shying away from the pole. The pole straps are fastened as soon as the animals are backed into the pole. Notice that in double harness breeching is unnecessary as the animal holds the vehicle back by means of the pole straps.

* If Kay collar is used it must be put on over the horse's head upside down and turned around at the throat before being pushed back into position on the shoulders.

ENTERING THE VEHICLE.

Standing on the off side the driver takes the reins out of the terret and holds them in the right hand at about the length they will be when he is in the seat. The whip is taken up at the same time and also held in the right hand. The driver then steps into the vehicle, placing his right foot on the step, left foot on the floor of the vehicle and supporting himself with the right hand on the dashboard, the left on the seat. Note, the driver always sits on the right side of the vehicle, even though, with our traffic which keeps to the right, to sit on the left would be more logical. If a four-wheeled vehicle is being used the wheels should be "cut" slightly to the left to give the driver more room for entering. If a passenger is to be carried the wheels are then cut to the right so that he can mount on the left.

METHODS OF HOLDING THE REINS.

There are two recognised methods of holding the reins, the first is the "orthodox" method in which the reins are held in the left hand at all times, the near rein entering the hand over the forefinger and under the thumb, the off rein entering the hand between the second and ring fingers, the bight of both reins passing out of the hand below the little finger.

The advantage of this method is that it leaves the right hand free to hold the whip and to assist the other hand in turning, shortening the reins and stopping.

The second, or "modern" method is smarter looking and is most often used in shows. Here both hands are employed to hold the reins, the near in the left hand and the off as well as the whip in the right. The bight forms a loop running between the hands.

SHORTENING THE REINS.

If the reins are held by the modern method the off rein must first be transferred to the left hand, then the right hand is placed in front of the left and the reins are pushed back through the latter. It is considered bad form to pull the reins through the fingers from behind.

STARTING.

The driver having taken his place and adjusted his reins and whip, he feels the horse's mouth to get his attention and then starts him off with a word or oral signal. The animal should not start with a jerk, but should move off slowly, not taking his regulation fast pace for several steps.

TURNING.

If the reins are held in the modern manner, the turn is made by a direct pull on one or other rein. If the orthodox method is used the right hand assists the left by shortening the desired rein from the front.

HILLS.

Unless the hill is very steep or slippery the horse will be more comfortable if allowed to trot up and down it. If the vehicle is equipped with a brake it may be used but should be used as sparingly as possible as it is better for the horses to become accustomed to holding the vehicle back themselves. If the hill has soft or sandy shoulders, the animal will find it easier to hold back if the right wheel is allowed to run on the shoulder.

PACE.

Except for the first few minutes when it is advisable to let the horse warm up slowly, and at the end of the drive when it may be necessary to cool him out, it is best for the horse to be kept at as fast a trot as is natural to him to take.

STOPPING.

The horse is stopped by the following method. If the reins are held in both hands the near rein must be transferred to the left hand. The right hand is then placed on the reins six inches to a foot or more in front of the left hand, grasping the reins at this point. The right hand then pulls back on the reins, at the same time the left hand is raised and brought forward until it is above the right and directly over it. Never attempt to stop by simply pulling back on the reins with one or both hands, when your hands reach your body you will have "nowhere to go" and if the horse has not already stopped you will have no further control over him. Do not bring the horse to an abrupt halt, but allow him to slacken his pace slowly, coming to a standstill at the desired point. Neither should you bring him down to a walk several yards from where the vehicle is to be stopped. The brake is applied only after the vehicle is at a standstill.

USE OF THE WHIP.

The whip is used for three purposes, to urge the horse forward, to correct him for misdemeanors (shying, biting, kicking, etc.) and to signal to other vehicles your intentions as to turning, etc.

The whip should *never* under any circumstances, be used when the rein is held in the right hand. This invites disaster as in bringing the hand forward to apply the whip the rein is inadvertently loosened and the horse is at liberty to plunge forward. It is nearly always necessary to steady the horse with both reins at the time when the whip is applied, especially when it is being used for correction.

The whip is used much more frequently in driving a pair than in driving a single horse inasmuch one of the two will often try to avoid work by slacking his traces. It then becomes necessary to touch him and make him come into his collar. In four-in-hand and tandem driving the driver depends entirely on his whip to control his leaders inasmuch as they can get away from the discipline of the reins by slackening their traces. It goes without saying that the horse should never be slapped with the reins.

DRIVING A PAIR.

The orthodox method of holding the reins is vastly to be preferred in driving a pair inasmuch as the frequent use of the whip may be imperative. Turns may be made in the following manner: Fold back the desired rein so that it forms a loop a few inches long which is held in place by the thumb. After the turn is made the loop is released when the reins will be automatically returned to their original lengths and the horses will straighten themselves out. By using this method the right hand is left free to use the whip on the turns and there is no danger of one rein becoming shorter than the other. The procedure is known as making a "point" and is also used in four-in-hand and tandem driving.

CHAPTER X

Starting the Child

THE PARENT AS TEACHER.

As a general rule, parents make poor teachers, not only becaus of their inexperience in the art of teaching, and consequently the ignorance as to how much a child can absorb, how fast he shoul progress, how to analyse his faults and correct them, but also be cause of the personal relationship between parent and child an the emotional reactions which result from this.

No man is a hero to his valet, likewise, children will accept a facts statements from strangers, but will question and often refus to accept the same statements from a parent. There is also the prob lem of the perhaps unconscious desire of the parent that the chil shall not only make normal progress but that he shall do better tha little Johnny who lives next door, even though Johnny may hav the benefit of age, temperament or experience. One sees this sam reaction in every-day life. If your neighbour's child comes in wit a dirty face when there is company for tea, you shrug your shou ders and think, "oh, well, little boys of that age are always dirty, but if your own offspring makes such an appearance you are horr fied and humiliated and banish him quickly from the public gaz This is one of the most difficult things to overcome in the parer who teaches his own child, or even in the one who simply come to watch the class. Observe such a parent, a good horseman himse. and naturally interested in the progress of his son or daughter. H watches the class with an intense and critical eye. Seeing Sonn hoist himself up and down by means of his hands and a steady gr on the reins is agony to him he fails to notice that every other chil in the ring, being at about the same stage in the careers as equestrians are doing the same or worse! In his opinion it is only Sonny who is awkward and something should be done about it immediately.

What is the effect of such an attitude on the young rider? To begin with it makes him feel that nothing that he does is right, consequently that he is failing, and that the task of learning to ride is both hard and disagreeable. One of the most difficult things to teach a beginner, especially one with poor co-ordination or who is, by nature, somewhat hesitant about new adventures, is relaxation. The parent who is constantly correcting, constantly seeking to teach his pupil everything there is to know about riding in one easy lesson (or even in one month of lessons), makes relaxation an impossibility.

Furthermore the parent has forgotten many of the difficulties which he himself had when he was a beginner. He does not anticipate the problems to be met and show his child how to handle them, he waits until the unexpected has happened, the undesired occurred.

An example of this is the tendency of a pony or horse to quicken his gait when the stable is in sight. The pupil may have been doing very well, he may be posting, as long as the pace is kept down to the jog trot, and the proud parent is more than pleased. Then the stable comes in sight, the pony lengthens his stride, the child, caught off balance, grabs for the pommel and pulls up his legs. The animal, realising by the slack rein that no one is in control, and further spurred by the pressure of the child's heels, trots faster or even breaks into a canter, taking his young rider not only into the stable but possibly right into the stall itself!

Such an experience is far more terrifying to a beginner than that of an ordinary tumble. It makes him realise only too well, that he is entirely at the mercy of his mount. That this is undoubtedly the fact is something that the teacher wishes to keep under cover, not only as far as the pupil is concerned but also as far as the horse is concerned for the mutual benefit of both.

What would the experienced instructor have done in a like circumstance? First, before coming within sight of the stable he would have slowed down to a walk, putting his own horse a little in advance and to the left of his pupil. Thus if the pony tries to increase his gait it will be an easy matter to block him off. Secondly, he would have explained to the child that horses do like to get home fast to supper, just as boys and girls do, and so it is just as well to be prepared with a short rein, just in case.

The over-cautious parent should never attempt to teach his own child, nor should he watch the lesson lest his fears be transferred to the child making the instructor's job of instilling confidence even more difficult.

Then there is the parent who grew up on the ranch or farm and learned to ride before he can remember. He is the one who buys his child a horse and expects him to learn to ride with no further instruction. He forgets that the whole background is different. That the things which he absorbed, practically with his mother's milk so that riding came to be as natural to him as walking, are new and untried territory. No greater error could be made, no method of instruction so hopeless as this one. Rare indeed is the child who has the nerve and the perseverance to overcome such a start and become a good horseman. Even if he manages to learn to stay on his horse and to control him, he will contract such bad habits that he will probably never learn to ride correctly, never be able to handle a sensitive, spirited horse so that both he and his mount are comfortable.

But if the picture of the average parent as teacher is a black one, there is the compensation that if you *can* forget your relationship to your child to the extent that he is merely one that is coming to you to learn to ride, and for whom you have only the normal ambitions and standards, and if you have sufficient experience in teaching, seeing your own child develop under your tutelage will give you a vastly greater amount of pleasure and satisfaction than the average instructor who works only with the children of strangers, experiences. And your child, able to have more frequent lessons, will progress faster.

It would be wise to try and take on the child of a neighbour as well as your own, even though it means that the children must take turns. Thus you will have some means of comparison, the children will enjoy their lessons more than if one were alone, and they will have more confidence as each will see that what he finds difficult is likewise difficult for the other. The clever teacher, finding that one child does better at certain things while the second excels in others, will use these differences to build confidence in the timid child and to teach the bold one that simply to be able to stick on is not all there is to riding.

GROUP INSTRUCTION VERSUS PRIVATE LESSONS.

Very young beginners do best in groups of not more than four or six riders. Intermediate and older children will make more progress and enjoy themselves more if they ride with twelve or more others who are of a similar age and experience. Only the child who is preparing for the horseshow and needs a little extra polishing and attention to fine points needs or enjoys private instruction. He will benefit by an hour or so alone.

The above is true, not only for the reasons already given, but because through games, drills and competitions of one sort or another it is much easier to keep a group of children interested and relaxed than it is to hold the attention of one or two children. This is particularly true when the child has passed the initial stage of riding at which time just the fact of being on a horse is sufficient excitement, and has reached the stage where he must be induced to correct faults of posture, hands and seat, learning at the same time to manage his mount and be the engineer and not the passenger. He is still unable to go on to the more difficult feats such as jumping which again in themselves will be adventurous and will supply the needed stimulation.

When your child has learned to be a fair or even a good rider, don't expect him to keep up his interest unless he has some one with whom to ride, or some goal such as breaking a colt or training a horse for the show ring.

AGE AT WHICH THE CHILD LEARNS MOST READILY.

It is difficult to give any exact answer to this question, since it depends so much on the type of child, the type of mount and the type of riding that is being taught. If you yourself ride and are going to be the teacher, getting your child a quiet pony as soon as he is able to sit up and take notice is ideal. The sight of the pony will become as familiar and is taken for granted as the sight of a piece of furniture. Fear will never enter into the picture.

But don't expect your child to manage even the most docile creature without help. Put him on every day and take him for little walks, talking about the scenery, the happenings of the day, etc. Only occasionally should you admonish him to sit a little straighter or to keep his hands and heels down. Adjust your rides so that the child is reluctant to get off at the end. Set a definite goal when you start out. Make everything as easy as you can. Give the suppling exercises, suggested in a previous chapter, before you teach him to trot. When you do reach the latter stage, trot only a few steps at a time. The best way to lead a young child on a pony is to walk on the left, your left hand on the cheek-strap, your right on his knee. If you put one hand behind his back, as many people do, the child will tend to lean against your arm for support, and will miss it when you take it away. By pressing on his knee you steady him and at the same time teach him to keep his knees in. A child as young as two can learn to balance at a walk and trot and even to post, but the instructor must do all the controlling of the animal. Children under six, unless they have a great deal of practice are unable to co-ordinate mind and body quickly enough to manage even a very quiet animal.

From seven to ten or twelve the average boy or girl learns to ride very readily. How fast he progresses will depend largely on his own ability. The problem of fear, or at least of uncertainty will probably have to be overcome as the horse or pony will represent something entirely new and outside their field of experience. Even though the young rider may seem slow, timid or disinterested, don't despair. Many a child that has appeared very nearly hopeless at the start has come to love riding and to become just as skilful as his more venturesome brother. This type of child should be taken slowly and quietly. He should not be pushed ahead of his ability, and above all should not be allowed to give up because of slight discouragements. Some children take as long as six months of lessons to learn to relax and enjoy themselves on a horse. It is very often this type of child who, in the end, enjoys its most, because usually he is the type who has few other skills.

The "teen-age" is a bad time to start a youngster, be it a boy or a girl. In the case of the former, football, baseball, etc. hold his interest and it is hard to get him to enjoy a sport that offers little excitement until the rider has developed a fair amount of skill. Girls at this age have poor co-ordination and balance. They are self-conscious and timid, learning far more slowly than their younger sisters.

It is seldom that a woman, taking up riding after she is grown, becomes really expert. However she can reach the intermediate grade and will enjoy herself on a well-trained and not too experienced horse. Naturally there are exceptions to this rule, but few women have the courage or the perseverance to tackle the more strenuous exercises which give the necessary seat and balance.

On the other hand boys of eighteen or more, entering military service often make bold and splendid horsemen even though they may never have ridden previously. This is because, want to or not, they are put through a strenuous course of training under discipline. Perhaps, in like circumstances, women would do as well.

PSYCHOLOGICAL PROBLEMS TO BE SOLVED.

In learning to ride the child is faced with two of the fundamental fears of mankind; the fear of falling and the fear of the unknown. In prehistoric times our aboriginal forebears had to develop and maintain these two fears even while asleep or they would have speedily been lunch for the lions. Is it any wonder that the average child is cautious in his approach?

Neither fear can be conquered except through the medium of *successful experience.* The child must learn, not by being told but by trying and succeeding, exactly what to expect and how to manage himself and his mount when the unexpected occurs. Falls are to be avoided until the child has thoroughly mastered the art of vaulting off his horse in an emergency, see page 131. Once having learned to land properly, falls do no harm, in fact they go far to instilling confidence as the child learns that such an experience is little to be feared.

PROGRESS TO BE EXPECTED.

Most children taking outside instruction get an average of two hours a week of riding. If they have more their progress will naturally be more rapid. With a child under ten, taking two one-hour lessons or one two-hour lesson a week, the first year should be spent in developing in him a love of horses and a love of riding. If, in addition, he can learn to manage a quiet mount under ordinary circumstances, to sit with a fair degree of form at a walk and a trot, to canter bareback on a well-trained pony, then he has learned all that can be expected of him. Some few children will go beyond this the first year, but these are the standards which fit the average case.

During the second and sometimes the third year, depending on the age and ability of the child as well as the amount of time that he spends on his lessons, he should materially improve his form. He should learn to manage more difficult horses, to play simple mounted games, to do the easier manœuvres of military drill, to saddle, bridle, groom and care for his horse, to ride over rough ground without fear and up and down narrow and steep trails, to ride the more spirited and difficult ponies bareback and sometimes to start low jumping. The progress that a child makes at this period is sometimes not as spectacular as the change from the complete beginner to the early intermediate stages, and the parent may become discouraged, but all the time muscles are being developed and balance is improving.

From the third year on the child should become fearless at all gaits and over hurdles. He should be trained for showing in the horsemanship classes to develop in him an interest in form. He should be given the opportunity of schooling a horse or colt for some special purpose such as jumping or light dressage. If there is a hunt in the vicinity he should be ready for cubbing at least. A new sport in this line is the sport of hunting with bloodhounds, familiar in England but only recently introduced in this country. It combines the good points of a paper chase, a drag and a fox hunt. All that is needed is to buy a couple of bloodhound pups and train them for this purpose; a thing very easily done, for the bloodhound is gentle and teachable with the finest nose in the world.

The procedure is for one rider to ride off and follow any course that pleases him and that is suited to the country and to the abilities of the children and their mounts. He may circle, ride through water, do anything which he thinks of to confuse the hounds. When he comes to the end of his trail he hides.

Meanwhile the hounds are held until the "fox" has been given enough time to get well away, and then they are turned loose. They will pick up the scent of the rider and his mount at once, it is entirely unnecessary to lay any kind of trail by dragging a bag of anise or fox's litter along the ground, and will stick to it through thick and thin. Meanwhile neither the child who is acting as huntsman nor the rest of the field have any idea of where "Mr. Fox" has hidden himself, so they have all the excitement of watching hounds work out an invisible trail. One who has never participated in these hunts cannot conceive of the skill and pertinacity of these dogs as well as of their speed, for they run just as fast as it is comfortable to go and if you lose sight of them, even for a minute, it will be hard to catch up. When the "fox" is found the hounds are rewarded by a "worry," a juicy piece of meat, to encourage them to run well another day.

Field manœuvres and war games with all the close and open order drills also do a great deal to maintain interest and make good riders of the youngsters in their teens who have had three or more years of riding. The degree of skill to be attained is limited only by the ability of the individual.

IMPORTANCE AND CHOICE OF MOUNT.

If it is hard to find a suitable mount for the adult rider, one that will suit him in size, disposition, gaits and price, it is doubly so to find one for the child. Whether or not the latter likes and can manage his horse or pony will make the difference between whether or not he continues his riding.

After many years of teaching I am more firmly convinced than ever that for young children a tractable pony is the only answer. A six-year-old, or even a ten-year-old who is not experienced, both looks and feels uncomfortable on a horse no matter how quiet the latter may be. In addition to the added distance he is from the ground, making falls more serious than they should be, his legs, coming as they do very little if any distance below the skirts of the saddle, make the use of them as aids practically impossible.

Ponies have gotten a bad name due in part to their inherent stubborness and to the fact that the English type Shetland, with his broad back and thick neck, is not as a rule suitable for riding. But there are many other types of ponies described in an earlier chapter, and if you can find a well-trained, gentle animal your child will make far faster progress than would be possible on a horse. Furthermore he will learn control more readily and will be able to do the more strenuous riding such as the vaulting and the riding without stirrups.

Test a prospective pony thoroughly for bad stable manners such as nipping and for his willingness to leave the stable. As far as size goes, the old rule that if the soles of the rider's feet come on a line with the bottom of the horse's belly, the mount is the right size for the horseman, can be used as general measure but need not be adhered to very strictly.

Try and find a pony not younger than six or seven and not older than fifteen. Ponies live longer than horses and they are sturdier, but the pony that is too old is prone to stumble and is sometimes harder headed than the younger animal. A pony younger than six is seldom steady enough for the beginner.

GOAL TO BE ACHIEVED.

When the child looks as though he belonged on his horse and the horse looks equally comfortable; when he can take a difficult, nervous colt or a tough mouthed, wise animal that knows exactly how to take advantage, and can make both go well; when riding without reins and stirrups is as easy for him as walking; when, without outside suggestion or supervision he puts the comfort of the horse before his own, never mishandling him, taking time to cool him out properly, etc., when he knows how a horse should be cared for and what to do for simple illnesses and accidents; when he knows how to conduct himself on the hunt field, in the show ring and on the bridle-trail, then you may know that your task as teacher is accomplished. That were this rider not to have the opportunity of riding again for ten years or more, he would be able to go out again with no more than a little temporary stiffness to tell him that it had been some time since he had straddled a horse. Furthermore, no matter where he rode or on what he was mounted he would be both safe and comfortable.

CHAPTER XI

The Show Ring

Horse showing is no longer a pastime limited only to those who can afford the best in horses, there are many local shows and children's shows where the members of a community may get together to compete in a friendly fashion and where the performance of the rider or of the horse counts more than the conformation or quality of the animal.

At the National Horseshow and the other big shows the winner is still the horse that is trained and used for show purposes only, and whose owner can afford to spend the necessary time and money. There is but one thing to be deplored in horse showing, that the animals trained by professionals for open jumping are misused, being forced to jump higher and higher on pain of severe punishment via the rapping pole is only too true. If they were punished only when they failed either deliberately or through carelessness it would not be so bad, but one has only to watch one of these poor creatures being schooled, see him exert himself to the utmost, clearing almost impossible obstacles, only to receive the stinging blow of the pole on his belly in mid air. At a recent show the humane society confiscated a pole studded with nails which was being used in this manner. Surely no real horse lover would allow his animal to be so treated regardless of the reward in either glory or money. Horses trained by the army for competition in military classes are far differently treated, yet perform equally well, showing that there is no reason for the abuse.

READING THE CATALOGUE.

Many people who go to the shows would enjoy them more if they were more familiar with the rules governing each class. The programme contains the following information: Name of the class. Restrictions and qualifications of the entries. Height of jumps. Age or sex limitations of rider or driver. Time and date when the class will be held.

Then follows the name and abbreviated description of each entry with the name of its owner. The former refer to the colour, sex, height, age and registration number of the animal in that order. Colours are described as follows: bay, b., brown, br., black, bl., chestnut, ch., grey, gr., roan, r., piebald, p.b., skewbald, sk. The sex is abbreviated thus: mare, m., gelding, g., stallion, s. The height is given in hands and inches, thus 15:2½ means fifteen hands, two and a half inches. The age may be given in years or, if the horse is over eight years old, he may simply be termed "aged." The following, "BETSY LEE, br.m., 14:0½, aged, 17926," would be the description of a pony named Betsy Lee who was a brown mare, fourteen hands and a half inch tall, over eight years of age and whose registry number was 17926. This description would be preceded in the catalogue by a number which would be the one worn or carried by the exhibitor of Betsy Lee in this class and it would be followed by the name of the owner.

In addition to the rather enigmatical abbreviations, amateur spectators at shows are frequently puzzled by such phrases as "green hunter," "novice jumper," "amateurs to ride," etc. and as to the exact difference between a working hunter, a hunter hack and a qualified hunter. They do not know wherein the difference in penalties lies in judging hunter classes, open classes, touch-and-out, or scurries. The knowledge of these fine points will do much to clarify the decisions of the judges.

CLASSIFICATIONS OF HORSES IN THE VARIOUS CLASSES.

SADDLE CLASSES. The paragraph below the number of the class will tell you whether the class is open to horses or ponies. It may even be more specific such as a class "open to saddle ponies not under twelve hands or over fourteen two." It will say whether the horses are required to show three or five gaits, it may specify amateurs (persons who do not make their living by dealing in or showing horses), ladies or children only to ride.

The saddle horse should have high action and present a flashy, commanding appearance. He should show a flat-footed, brisk walk, an evenly cadenced, high trot and a very slow, collected canter. If in a five-gaited class he will also be asked to demonstrate a "slow-gait" that is rapid in movement, and a fast rack. In the "Ladies to Ride" classes, manners and tractability are all important. Professionals showing saddle horses often sit the trot, their bodies way back against the cantle of the saddle, legs stretched out in front of them. This is not due to a lack of knowledge of horsemanship, but is designed to demonstrate the smoothness of the animals gait and to show off his front. In reversing in a saddle class the horse is reversed towards the centre of the ring, the opposite of which is required in a horsemanship class. Saddle horses are asked to demonstrate all gaits travelling in both directions and to back. Some judges like to pick out the best horses first, lining them up in the middle of the ring while they consider the merits of the others. The latter are then excused while the picked animals continue to perform until a decision is reached. Others prefer to pick the poor ones out first, keeping them lined up in the ring until the judging is finished.

ROAD HACK CLASSES. These classes are for saddle horses, not necessarily of the flashy type, whose chief qualification is their speed at the trot. They are usually required to demonstrate the walk, the slow or collected trot, the fast or extended trot, the canter and the gallop as well as to back.

BRIDLEPATH HACK CLASSES. Such classes may be open either to horses of the saddle type or of the thoroughbred type. The exhibitors in these classes ride with as loose a rein as possible to show the gentleness and tractability of their mount. Sometimes a low jump is required or the ability to open and shut a gate. The horse that appears to give the most comfortable ride under ordinary hacking conditions is the one that wins. Conformation may not count at all or it may count twenty-five percent depending on the stipulations.

BRANDED HORSES. This class is for cow-ponies which must be branded. They are shown under western saddles and are required to demonstrate their speed at a gallop, ability to stop and turn quickly and general flexibility of handling.

PAIRS OF SADDLE HORSES. This class is judged on the appearance and performance of the horses as a pair, similarity counting very high. The requirements are the same as in SADDLE HORSE CLASSES.

Western Trappings

The saddles and bridles are silver mounted. Because of the severity of the bits the Westerner rides with a light hand and loose reins, the horses being taught to turn and stop the instant they feel the shift of weight or are given the signal of the bearing rein.

COMBINATION CLASSES. These classes are for horses or ponies that can be both ridden and driven. The animal enters the ring in harness, the saddle and riding bridle being in the vehicle. After demonstrating his ability as a harness horse, he is unhitched in the ring, and saddled up while the groom or assistant runs the vehicle out of the way.

RIDE, DRIVE AND JUMP CLASSES. These classes were introduced only a few years ago. After being shown in harness and under the saddle the best six or eight horses are asked to jump a three-foot hurdle to demonstrate their flexibility and usefulness.

HARNESS CLASSES. These classes may be restricted as to the type of vehicle to be used, to the light- or heavy-weight harness type. They may be limited as to the size of the entry or may be for ladies or children only to drive. The horses are asked to demonstrate smoothness and evenness of cadence at the trot, to back, walk and sometimes execute a figure eight. Many judges will not award a ribbon without first driving the entry themselves. The harness horses and ponies at the big shows are marvellous as to their quality and performance, but it is also fun to go to a local show and see the little girls in pigtails with their fat, slow little ponies, trotting around the ring just as proud as though their pet had all the action and quality in the world!

MAIDEN CLASS. Open to horses that have not won a blue ribbon in a recognised horseshow in the division in which they are being shown.

NOVICE CLASS. Open to horses that have not won three blue ribbons in a recognised horseshow in the division in which they are being shown.

LIMIT CLASS. Open to horses that have not won six blue ribbons in a recognised show in the division in which they are being shown.

Jumps for above classes do not exceed three feet six inches. In pair and team classes horses may enter Novice, Maiden and Limit classes even though they may have won more than the stipulated number of ribbons provided they have not done so when exhibited as a pair or team.

SUITABLE TO BECOME CLASSES.*

Open to horses four years old or under which have not won a first ribbon at any recognised show in the Hunter Division except in classes either requiring no jumping or in which the jumps do not exceed three feet, six inches.

* In this class the word "Hunter" is understood.

HUNTERS AND JUMPERS.

HUNTERS. Hunters are judged not only on their performance over the jumps, but also on their conformation, mannners and way of going. In out-door shows the hunter classes are shown over an outside course of jumps which simulate as far as possible actual conditions in the hunt field. There is usually plenty of room between the obstacles to give the entry room to maintain a steady hunting pace and to enable the judges to decide which of the animals would carry his rider most comfortably and safest in the hunt field and which would be most likely to give trouble.

The hunter should keep an even gallop, fast enough to keep up with hounds but in no way a racing pace. He should take his jumps in his stride, standing well back and going no higher than the jump requires. Ticks (light touches) are not usually scored in hunter classes. His manners, especially in the "ladies hunter" classes should be beyond reproach, he should show that he is sound of wind and limb, has a good mouth and is tractable.

After being ridden the animal is stripped (saddle taken off) and is judged for conformation. Hunters are classified as to their ability to carry weight rather than as to actual size. A light-weight hunter must be up to carrying a hundred and sixty pounds in the hunt field, a medium-weight up to two hundred and a heavy-weight more than two hundred. "Green hunters" are horses of the hunter type which have not hunted with a recognised pack for more than one season, or have not won a first ribbon in any classes other than the classes for green hunters where the jumps are limited to three feet, six inches.

QUALIFIED HUNTERS are horses that have been hunted for more than one season with a recognised pack as evidenced by a certificate from the Master of that pack.

HANDY HUNTER CLASSES. These are designed to show the flexibility and tractability of the horse, various tests being given. He may be asked to stand while the rider takes off the top rail of a jump and dismounts, the horse being then led over the obstacles. He must stand quietly for mounting. Or the rider may remove the bar from his saddle, rein the horse back a few feet and then require him to jump from this distance and without a run. He may be asked to walk over one jump, trot over the next and canter over the third.

Over an outside course he may be required to stop at the sound of the hunting horn, stand without excitement and then continue on his way. Performance usually counts more than conformation in such a class.

HUNTER HACKS are required to show that they are comfortable and easy to ride, going steadily on a loose rein. They must take low jumps without excitement and they may be asked to stand while the rider opens a gate or takes down a rail.

CORINTHIAN CLASSES. Classes limited to members of recognised hunts, wearing hunt livery and carrying personal and hunting equipment (sandwich cases and flasks, hunting thongs, string gloves under the billets and breastplates, the latter being optional.)

HUNT TEAM CLASSES. Classes for teams of three riders in hunting livery who ride one behind the other at a safe hunting distance. As a rule a light-, a middle-weight and a heavy-weight hunter compose the team, being matched as far as possible in colour and type. Manners, way of going, conformation, spacing, pace, and similarity all count.

WORKING HUNTER CLASSES. These are for hunters in which conformation and "honourable scars and blemishes" do not count. The horse is required to be "hunting sound" only. Performance, manners and way of going to count.

OPEN JUMPING CLASSES.

The difference between the above and the hunter classes lies in the fact that the open jumper is scored on his performance only, a system of penalties, see page 163, having been adopted by which each fault (touching the jump, falling, shying out, or refusing) counts a certain number of points against him. Manners, way of going and conformation do not count. Open classes are always held *inside* the ring. The horse is usually required to take eight jumps of a specified height set either along the walls or arranged in a special sequence. After every horse has jumped one round, the scores are tallied, and in cases of a tie, the tied horses are asked to jump again, the jumps having been raised. There is no limitation as to the size of the animal, one occasionally sees ponies of fourteen hands or under competing against sixteen- or seventeen-hand animals.

STAKE CLASSES. These are open classes in which the winner, instead of being awarded a trophy, receives a percentage or all of the entry fees plus any other money enumerated. Sometimes the money is divided between the three top horses.

A Thoroughbred Taking a "Hog's Back" in an Indoor Show

TOUCH-AND-OUT CLASSES. These differ from the regular open classes in that as soon as the horse touches the jump he is signalled out of the ring by a doleful toot on the horn of the ring master and is not allowed to complete the course. *Strip fillets,* light wooden laths are sometimes put on top of the topmost bar of the obstacle. These flutter to the ground as soon as touched and so prove without question that the horse has ticked. *Knock-down-and-out* classes are judged the same way except that here the entry is eliminated only if he actually knocks off a bar. One refusal in either class constitutes elimination. The winner is the animal which can get over the most jumps without a fault. In cases of ties the jumps are raised and the competition continues.

TRIPLE-BAR CLASSES. Here the obstacles are composed of three bars of graduating heights placed two or more feet apart, all three of which must be taken at a leap. In case of ties the bars are sometimes separated still further as well as raised. A knock down in a triple-bar class must affect the height of the highest bar only, knock downs of the lower bars not counting as much. Ladies are not permitted to ride in Triple-bar classes.

SCURRIES. The scurry is a class in which the time which the rider takes to complete the course is the only thing which counts, one second being added for each fault.

MILITARY JUMPING CLASSES. These classes are open only to members of a military organisation who must ride in uniform. The scoring is different from that of the international horseshow scoring. The jumps are usually arranged in a somewhat complicated course and the obstacles are very varied in character.

PAIR JUMPING. In these classes the horses jump in pairs abreast, they are judged on appearance and performance as a pair. In the event of both horses having faults over the same jump, only the major fault is counted.

HORSEMANSHIP CLASSES.

These classes are usually for children, the ages or experience of the children being specified in the programme. Very young children are sometimes shown on a lead rein and only asked to walk and trot. Intermediate riders may be asked to walk, trot, canter and back. Experienced riders may also be required to demonstrate their skill by such things as performing a figure eight, changing the lead, taking a designated lead on a straight line, taking the wrong lead, etc.

HORSEMANSHIP JUMPING CLASSES usually allow any child under eighteen to compete. In local shows the jumps are usually three feet in height, in large shows, three feet six. Faults, unless they are the fault of the rider, are not counted.

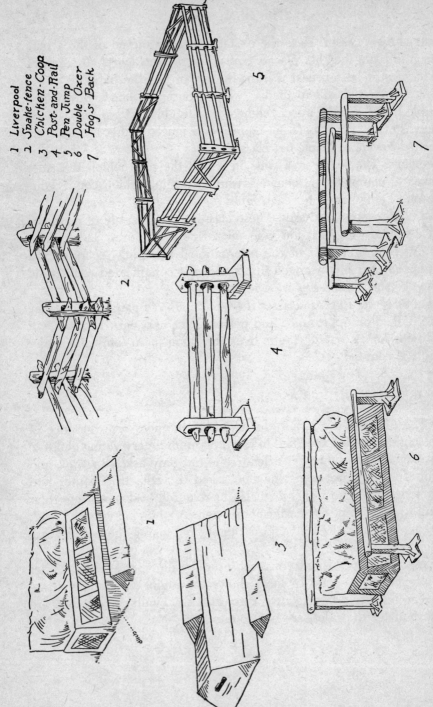

1 Liverpool
2 Snake-fence
3 Chicken-Coop
4 Post-and-Rail
5 Pen Jump
6 Double Oxer
7 Hog's Back

Types of Jumps Used in Hunter and Jumper Classes

TYPES OF JUMPS.

POST-AND-RAIL. A simple fence of three or more bars.

SNAKE-FENCE. Angular type of rail fence found in the south.

OXER. A hedge with a rail on the near, or take-off side, a double oxer has a rail on both sides.

BRUSH JUMP. A hedge made of growing shrubs or of evergreens in a wooden base.

HOG'S BACK. An obstacle consisting of three poles, the middle one being higher than the others.

BULL FINCH. A hedge jump with an opening midway up through which the rider jumps.

BANK JUMP. An obstacle built in such a way that the rider must either land on top of it or jump off of it, if the latter it is called a "drop jump."

CHICKEN COOP. An inverted V-shaped wooden obstacle either weathered or painted.

AIKEN FENCE. An obstacle made of light-weight fir-tree poles.

LIVERPOOL. An obstacle consisting of take off bar, ditch and hedge.

IN-AND-OUT. An obstacle consisting of two jumps set about twenty-four feet apart so that the horse must take off for the second immediately after landing from the first.

PEN JUMP. An obstacle made in the form of a pen. The horse may be required either to jump in and then out over the opposite panel, or he may be asked to jump in, turn and jump out at right angles.

PICKET FENCE AND GATE JUMPS. Require no explanation.

SCORING OF HUNTERS AND JUMPERS.

Touch with any part of the body behind stifle	½ fault
Touch with any part of body in front of stifle	1 fault
Touch of obstacle standard or wing	1 fault
Knock-down with any part of body behind stifle*	3 faults
Knock-down with any part of body in front of stifle	4 faults
Knock-down of obstacle by rider	4 faults
Horse and/or rider falling †	debarred

* An obstacle is considered knocked down when its height is lowered by horse or rider.

† A horse is considered to have fallen when the shoulder and haunch on the same side touch the ground.

First refusal, run-out or bolting off course.............. 3 faults
Second refusal .... 6 faults
Third refusal .... debarred
Jumping an obstacle before it has been reset........... debarred
Horse bolting from ring, mounted or riderless.......... debarred
Failure to keep proper course........................ debarred
Circling a horse between obstacles .... 1 fault

HORSE SHOW AWARDS.

In all except the stake and other money classes the awards are as follows:

1st Prize	....	Trophy and Blue ribbon
2d "	....	Red ribbon
3d "	....	Yellow ribbon
4th "	....	White ribbon
5th "	....	Pink ribbon
6th "	....	Green ribbon
7th "	....	Purple ribbon
8th "	....	Brown ribbon

Grand Champion ....Blue, Red, Yellow and White
Reserve to Grand Champion..Red, Yellow, White and Pink*
Champion ....Blue, Red and Yellow
Reserve to Champion ....Red, Yellow and White

Except in Horsemanship classes as a rule only the first four ribbons and the trophy are presented.

* The practice of choosing a Reserve to Champion was originated and designed to supply a second choice if the veterinary should declare the horse chosen as champion to be unsound.

Glossary

Aged—a horse more than eight years of age.

Barn rat—a horse that refuses to leave the stable.

Child-broke—a horse that is unusually gentle, accustomed to children and that can be handled safely by them.

Combination—a horse that will both ride and drive.

Clever—a dealer's term meaning very gentle and tractable.

Colors of a horse are as follows, given beginning with the darkest— Black, brown, bay, liver chestnut, chestnut (sorrel) and grey. All white horses except a very few are born black and turn white with age, hence they are all properly known as greys. The term roan designates a horse of a solid color with a mixture of white hairs through its coat. A piebald is a black and white spotted horse, a skewbald is brown or bay and white, the term "pinto" is Mexican meaning paint and refers to a spotted horse; a "paint pony" means the same.

Dog—a horse that will only move if urged with whip and spur.

Green Horse—one that has not had very much training. In the East it usually means a horse just shipped in from the West. For "Green Hunter or Jumper" see page 158.

Glass eyed, or moon eyed—one or both eyes light blue by nature, does not affect sight but detracts from appearance, common in spotted horses.

Hand—measurement of a horse, four inches, taken at top of withers

Headshy—frightened of being handled about the head.

Manshy—nervous about people, sometimes dangerously so.

Near side—left side of horse facing front.

Off side—right side of horse.

Stall courage—bucking or otherwise cutting up due to lack of exercise.

Horse—a mature male specimen of the Equine race, over fourteen hands two inches, other classifications as follows: A foal, a newly born horse; a colt, a male horse under four years; a filly, a female under four years; a stallion, a mature male; a gelding, a male that has been castrated to make him more tractable; a mare, a mature female; a pony, a member of the Equine race under fourteen hands two inches.

Star Gazer—horse that holds his head too high with his nose out especially over jumps.

To favor—to limp slightly.

To go short—to take short steps indicating soreness or slight lameness.

To be policed—to be thrown.
To tack up—to put on saddle and bridle.

Index

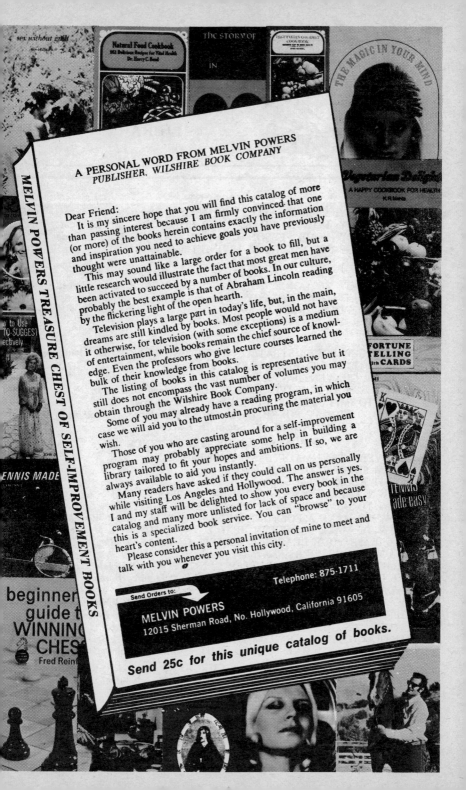

MELVIN POWERS TREASURE CHEST OF SELF-IMPROVEMENT BOOKS

Melvin Powers
SELF-IMPROVEMENT
LIBRARY

Melvin Powers
SELF-IMPROVEMENT
LIBRARY

_____ CHESS SECRETS REVEALED *Fred Reinfeld* 2.00
_____ CHESS STRATEGY — An Expert's Guide *Fred Reinfeld* 2.00
_____ CHESS TACTICS FOR BEGINNERS *edited by Fred Reinfeld* 2.00
_____ CHESS THEORY & PRACTICE *Morry & Mitchell* 2.00
_____ HOW TO WIN AT CHECKERS *Fred Reinfeld* 2.00
_____ 1001 BRILLIANT WAYS TO CHECKMATE *Fred Reinfeld* 2.00
_____ 1001 WINNING CHESS SACRIFICES & COMBINATIONS *Fred Reinfeld* 3.00

COOKERY & HERBS

_____ CULPEPER'S HERBAL REMEDIES *Dr. Nicholas Culpeper* 2.00
_____ FAST GOURMET COOKBOOK *Poppy Cannon* 2.50
_____ HEALING POWER OF HERBS *May Bethel* 2.00
_____ HERB HANDBOOK *Dawn MacLeod* 2.00
_____ HERBS FOR COOKING AND HEALING *Dr. Donald Law* 2.00
_____ HERBS FOR HEALTH How to Grow & Use Them *Louise Evans Doole* 2.00
_____ HOME GARDEN COOKBOOK Delicious Natural Food Recipes *Ken Kraft* 3.00
_____ MEDICAL HERBALIST *edited by Dr. J. R. Yemm* 3.00
_____ NATURAL FOOD COOKBOOK *Dr. Harry C. Bond* 2.00
_____ NATURE'S MEDICINES *Richard Lucas* 2.00
_____ VEGETABLE GARDENING FOR BEGINNERS *Hugh Wiberg* 2.00
_____ VEGETABLES FOR TODAY'S GARDENS *R. Milton Carleton* 2.00
_____ VEGETARIAN COOKERY *Janet Walker* 2.00
_____ VEGETARIAN COOKING MADE EASY & DELECTABLE *Veronica Vezza* 2.00
_____ VEGETARIAN DELIGHTS — A Happy Cookbook for Health *K. R. Mehta* 2.00
_____ VEGETARIAN GOURMET COOKBOOK *Joyce McKinnel* 2.00

HEALTH

_____ DR. LINDNER'S SPECIAL WEIGHT CONTROL METHOD 1.00
_____ GAYELORD HAUSER'S NEW GUIDE TO INTELLIGENT REDUCING 3.00
_____ HELP YOURSELF TO BETTER SIGHT *Margaret Darst Corbett* 2.00
_____ HOW TO IMPROVE YOUR VISION *Dr. Robert A. Kraskin* 2.00
_____ HOW YOU CAN STOP SMOKING PERMANENTLY *Ernest Caldwell* 2.00
_____ LSD — THE AGE OF MIND *Bernard Roseman* 2.00
_____ MIND OVER PLATTER *Peter G. Lindner, M.D.* 2.00
_____ NEW CARBOHYDRATE DIET COUNTER *Patti Lopez-Pereira* 1.00
_____ PSYCHEDELIC ECSTASY *William Marshall & Gilbert W. Taylor* 2.00
_____ YOU CAN LEARN TO RELAX *Dr. Samuel Gutwirth* 2.00
_____ YOUR ALLERGY—What To Do About It *Allan Knight, M.D.* 2.00

HOBBIES

_____ BLACKSTONE'S MODERN CARD TRICKS *Harry Blackstone* 2.00
_____ BLACKSTONE'S SECRETS OF MAGIC *Harry Blackstone* 2.00
_____ COIN COLLECTING FOR BEGINNERS *Burton Hobson & Fred Reinfeld* 2.00
_____ 400 FASCINATING MAGIC TRICKS YOU CAN DO *Howard Thurston* 3.00
_____ GOULD'S GOLD & SILVER GUIDE TO COINS *Maurice Gould* 2.00
_____ HOW I TURN JUNK INTO FUN AND PROFIT *Sari* 3.00
_____ HOW TO WRITE A HIT SONG & SELL IT *Tommy Boyce* 7.00
_____ JUGGLING MADE EASY *Rudolf Dittrich* 2.00
_____ MAGIC MADE EASY *Byron Wels* 2.00

_____ SEW SIMPLY, SEW RIGHT *Mini Rhea & F. Leighton*		2.00
_____ STAMP COLLECTING FOR BEGINNERS *Burton Hobson*		2.00
_____ STAMP COLLECTING FOR FUN & PROFIT *Frank Cetin*		2.00

HORSE PLAYERS' WINNING GUIDES

_____ BETTING HORSES TO WIN *Les Conklin*	2.00
_____ HOW TO PICK WINNING HORSES *Bob McKnight*	2.00
_____ HOW TO WIN AT THE RACES *Sam (The Genius) Lewin*	2.00
_____ HOW YOU CAN BEAT THE RACES *Jack Kavanagh*	2.00
_____ MAKING MONEY AT THE RACES *David Barr*	2.00
_____ PAYDAY AT THE RACES *Les Conklin*	2.00
_____ SMART HANDICAPPING MADE EASY *William Bauman*	2.00

HYPNOTISM

_____ ADVANCED TECHNIQUES OF HYPNOSIS *Melvin Powers*	1.00
_____ CHILDBIRTH WITH HYPNOSIS *William S. Kroger, M.D.*	2.00
_____ HOW TO SOLVE YOUR SEX PROBLEMS	
WITH SELF-HYPNOSIS *Frank S. Caprio, M.D.*	2.00
_____ HOW TO STOP SMOKING THRU SELF-HYPNOSIS *Leslie M. LeCron*	2.00
_____ HOW TO USE AUTO-SUGGESTION EFFECTIVELY *John Duckworth*	2.00
_____ HOW YOU CAN BOWL BETTER USING SELF-HYPNOSIS *Jack Heise*	2.00
_____ HOW YOU CAN PLAY BETTER GOLF USING SELF-HYPNOSIS *Heise*	2.00
_____ HYPNOSIS AND SELF-HYPNOSIS *Bernard Hollander, M.D.*	2.00
_____ HYPNOTISM *(Originally published in 1893) Carl Sextus*	3.00
_____ HYPNOTISM & PSYCHIC PHENOMENA *Simeon Edmunds*	2.00
_____ HYPNOTISM MADE EASY *Dr. Ralph Winn*	2.00
_____ HYPNOTISM MADE PRACTICAL *Louis Orton*	2.00
_____ HYPNOTISM REVEALED *Melvin Powers*	1.00
_____ HYPNOTISM TODAY *Leslie LeCron & Jean Bordeaux, Ph.D.*	2.00
_____ MODERN HYPNOSIS *Lesley Kuhn & Salvatore Russo, Ph.D.*	3.00
_____ NEW CONCEPTS OF HYPNOSIS *Bernard C. Gindes, M.D.*	3.00
_____ POST-HYPNOTIC INSTRUCTIONS *Arnold Furst*	2.00
How to give post-hypnotic suggestions for therapeutic purposes.	
_____ PRACTICAL GUIDE TO SELF-HYPNOSIS *Melvin Powers*	2.00
_____ PRACTICAL HYPNOTISM *Philip Magonet, M.D.*	2.00
_____ SECRETS OF HYPNOTISM *S. J. Van Pelt, M.D.*	3.00
_____ SELF-HYPNOSIS *Paul Adams*	2.00
_____ SELF-HYPNOSIS Its Theory, Technique & Application *Melvin Powers*	2.00
_____ SELF-HYPNOSIS A Conditioned-Response Technique *Laurance Sparks*	3.00
_____ THERAPY THROUGH HYPNOSIS *edited by Raphael H. Rhodes*	3.00

JUDAICA

_____ HOW TO LIVE A RICHER & FULLER LIFE *Rabbi Edgar F. Magnin*	2.00
_____ MODERN ISRAEL *Lily Edelman*	2.00
_____ OUR JEWISH HERITAGE *Rabbi Alfred Wolf & Joseph Gaer*	2.00
_____ ROMANCE OF HASSIDISM *Jacob S. Minkin*	2.50
_____ SERVICE OF THE HEART *Evelyn Garfield, Ph.D.*	3.00
_____ STORY OF ISRAEL IN COINS *Jean & Maurice Gould*	2.00
_____ STORY OF ISRAEL IN STAMPS *Maxim & Gabriel Shamir*	1.00
_____ TONGUE OF THE PROPHETS *Robert St. John*	3.00
_____ TREASURY OF COMFORT *edited by Rabbi Sidney Greenberg*	3.00

MARRIAGE, SEX & PARENTHOOD

_____ ABILITY TO LOVE *Dr. Allan Fromme*	3.00
_____ ENCYCLOPEDIA OF MODERN SEX & LOVE TECHNIQUES *Macandrew*	3.00
_____ GUIDE TO SUCCESSFUL MARRIAGE *Drs. Albert Ellis & Robert Harper*	3.00
_____ HOW TO RAISE AN EMOTIONALLY HEALTHY, HAPPY CHILD, *A. Ellis*	2.00
_____ IMPOTENCE & FRIGIDITY *Edwin W. Hirsch, M.D.*	2.00
_____ NEW APPROACHES TO SEX IN MARRIAGE *John E. Eichenlaub, M.D.*	2.00
_____ SEX WITHOUT GUILT *Albert Ellis, Ph.D.*	2.00
_____ SEXUALLY ADEQUATE FEMALE *Frank S. Caprio, M.D.*	2.00
_____ SEXUALLY ADEQUATE MALE *Frank S. Caprio, M.D.*	2.00
_____ YOUR FIRST YEAR OF MARRIAGE *Dr. Tom McGinnis*	2.00

METAPHYSICS & OCCULT

_____BOOK OF TALISMANS, AMULETS & ZODIACAL GEMS *William Pavitt* 3.00
_____CONCENTRATION—A Guide to Mental Mastery *Mouni Sadhu* 3.00
_____DREAMS & OMENS REVEALED *Fred Gettings* 2.00
_____EXTRASENSORY PERCEPTION *Simeon Edmunds* 2.00
_____FORTUNE TELLING WITH CARDS *P. Foli* 2.00
_____HANDWRITING ANALYSIS MADE EASY *John Marley* 2.00
_____HANDWRITING TELLS *Nadya Olyanova* 3.00
_____HOW TO UNDERSTAND YOUR DREAMS *Geoffrey A. Dudley* 2.00
_____ILLUSTRATED YOGA *William Zorn* 2.00
_____IN DAYS OF GREAT PEACE *Mouni Sadhu* 2.00
_____KING SOLOMON'S TEMPLE IN THE MASONIC TRADITION *Alex Horne* 5.00
_____MAGICIAN — His training and work *W. E. Butler* 2.00
_____MEDITATION *Mouni Sadhu* 3.00
_____MODERN NUMEROLOGY *Morris C. Goodman* 2.00
_____NUMEROLOGY—ITS FACTS AND SECRETS *Ariel Yvon Taylor* 2.00
_____PALMISTRY MADE EASY *Fred Gettings* 2.00
_____PALMISTRY MADE PRACTICAL *Elizabeth Daniels Squire* 3.00
_____PALMISTRY SECRETS REVEALED *Henry Frith* 2.00
_____PRACTICAL YOGA *Ernest Wood* 3.00
_____PROPHECY IN OUR TIME *Martin Ebon* 2.50
_____PSYCHOLOGY OF HANDWRITING *Nadya Olyanova* 2.00
_____SEEING INTO THE FUTURE *Harvey Day* 2.00
_____SUPERSTITION — Are you superstitious? *Eric Maple* 2.00
_____TAROT *Mouni Sadhu* 4.00
_____TAROT OF THE BOHEMIANS *Papus* 3.00
_____TEST YOUR ESP *Martin Ebon* 2.00
_____WAYS TO SELF-REALIZATION *Mouni Sadhu* 2.00
_____WITCHCRAFT, MAGIC & OCCULTISM—A Fascinating History *W. B. Crow* 3.00
_____WITCHCRAFT — THE SIXTH SENSE *Justine Glass* 2.00
_____WORLD OF PSYCHIC RESEARCH *Hereward Carrington* 2.00
_____YOU CAN ANALYZE HANDWRITING *Robert Holder* 2.00

SELF-HELP & INSPIRATIONAL

_____CYBERNETICS WITHIN US *Y. Saparina* 3.00
_____DAILY POWER FOR JOYFUL LIVING *Dr. Donald Curtis* 2.00
_____DOCTOR PSYCHO-CYBERNETICS *Maxwell Maltz, M.D.* 3.00
_____DYNAMIC THINKING *Melvin Powers* 1.00
_____GREATEST POWER IN THE UNIVERSE *U. S. Andersen* 4.00
_____GROW RICH WHILE YOU SLEEP *Ben Sweetland* 2.00
_____GROWTH THROUGH REASON *Albert Ellis, Ph.D.* 3.00
_____GUIDE TO DEVELOPING YOUR POTENTIAL *Herbert A. Otto, Ph.D.* 3.00
_____GUIDE TO LIVING IN BALANCE *Frank S. Caprio, M.D.* 2.00
_____GUIDE TO RATIONAL LIVING *Albert Ellis, Ph.D. & R. Harper, Ph.D.* 3.00
_____HELPING YOURSELF WITH APPLIED PSYCHOLOGY *R. Henderson* 2.00
_____HELPING YOURSELF WITH PSYCHIATRY *Frank S. Caprio, M.D.* 2.00
_____HOW TO ATTRACT GOOD LUCK *A. H. Z. Carr* 2.00
_____HOW TO CONTROL YOUR DESTINY *Norvell* 2.00
_____HOW TO DEVELOP A WINNING PERSONALITY *Martin Panzer* 3.00
_____HOW TO DEVELOP AN EXCEPTIONAL MEMORY *Young & Gibson* 3.00
_____HOW TO OVERCOME YOUR FEARS *M. P. Leahy, M.D.* 2.00
_____HOW YOU CAN HAVE CONFIDENCE AND POWER *Les Giblin* 2.00
_____HUMAN PROBLEMS & HOW TO SOLVE THEM *Dr. Donald Curtis* 2.00
_____I WILL *Ben Sweetland* 2.00
_____LEFT-HANDED PEOPLE *Michael Barsley* 3.00
_____MAGIC IN YOUR MIND *U. S. Andersen* 3.00
_____MAGIC OF THINKING BIG *Dr. David J. Schwartz* 2.00
_____MAGIC POWER OF YOUR MIND *Walter M. Germain* 3.00
_____MENTAL POWER THRU SLEEP SUGGESTION *Melvin Powers* 1.00
_____ORIENTAL SECRETS OF GRACEFUL LIVING *Boye De Mente* 1.00
_____OUR TROUBLED SELVES *Dr. Allan Fromme* 3.00

A NEW GUIDE TO RATIONAL LIVING

Contents: *by Albert Ellis, Ph.D. & Robert A. Harper, Ph.D.*

1. How Far Can You Go With Self-Analysis? 2. You Feel the Way You Think 3. Feeling Well by Thinking Straight 4. How You Create Your Feelings 5. Thinking Yourself Out of Emotional Disturbances 6. Recognizing and Attacking Neurotic Behavior 7. Overcoming the Influences of the Past 8. Does Reason Always Prove Reasonable? 9. Refusing to Feel Desperately Unhappy 10. Tackling Dire Needs for Approval 11. Eradicating Dire Fears of Failure 12. How to Stop Blaming and Start Living 13. How to Feel Undepressed though Frustrated 14. Controlling Your Own Destiny 15. Conquering Anxiety 16. Acquiring Self-discipline 17. Rewriting Your Personal History 18. Accepting Reality 19. Overcoming Inertia and Getting Creatively Absorbed 20. Living Rationally in an Irrational World 21. Rational-Emotive Therapy or Rational Behavior Training Updated **256 Pages . . .$3**

SEX WITHOUT GUILT

Contents: *by Albert Ellis, Ph.D.*

1. New Light on Masturbation 2. Thoughts on Petting 3. On Premarital Sex Relations 4. Adultery: Pros & Cons 5. The Justification of Sex Without Love 6. Why Americans Are So Fearful of Sex 7. Adventures with Sex Censorship 8. How Males Contribute to Female Frigidity 9. Sexual Inadequacy in the Male 10. When Are We Going to Quit Stalling About Sex Education? 11. How American Women Are Driving American Males Into Homosexuality 12. Another Look at Sexual Abnormality 13. On the Myths About Love. 14. Sex Fascism 15. The Right to Sex Enjoyment. **190 Pages . . . $2**

A GUIDE TO SUCCESSFUL MARRIAGE

Contents: *by Albert Ellis, Ph.D. & Robert A. Harper, Ph.D.*

1. Modern Marriage: Hotbed of Neurosis 2. Factors Causing Marital Disturbance 3. Gauging Marital Compatibility 4. Problem Solving in Marriage 5. Can We Be Intelligent About Marriage? 6. Love or Infatuation? 7. To Marry or Not To Marry 8. Sexual Preparation for Marriage 9. Impotence in the Male 10. Frigidity in the Female 11. Sex "Excess" 12. Controlling Sex Impulses 13. Nonmonogamous Desires 14. Communication in Marriage 15. Children 16. In-laws 17. Marital Incompatibility Versus Neurosis 18. Divorce 19. Succeeding in Marriage 20. Directory of Marriage Counseling Services. **304 Pages . . . $3**

PSYCHO-CYBERNETICS

A New Technique for Using Your Subconscious Power

Contents: *by Maxwell Maltz, M.D., F.I.C.S.*

1. The Self Image: Your Key to a Better Life 2. Discovering the Success Mechanism within You 3. Imagination—The First Key to Your Success Mechanism 4. Dehypnotize Yourself from False Beliefs 5. How to Utilize the Power of Rational Thinking 6. Relax and Let Your Success Mechanism Work for You 7. You Can Acquire the Habit of Happiness 8. Ingredients of the Success-Type Personality and How to Acquire Them 9. The Failure Mechanism: How to Make It Work For You Instead of Against You 10. How to Remove Emotional Scars, or How to Give Yourself an Emotional Face Lift 11. How to Unlock Your Real Personality 12. Do-It-Yourself Tranquilizers That Bring Peace of Mind 13. How to Turn a Crisis into a Creative Opportunity 14. How to Get "That Winning Feeling" 15. More Years of Life—More Life in Your Years. **268 Pages . . . $2**

A PRACTICAL GUIDE TO SELF-HYPNOSIS

Contents: *by Melvin Powers*

1. What You Should Know About Self-Hypnosis 2. What About the Dangers of Hypnosis? 3. Is Hypnosis the Answer? 4. How Does Self-Hypnosis Work? 5. How to Arouse Yourself From the Self-Hypnotic State 6. How to Attain Self-Hypnosis 7. Deepening the Self-Hypnotic State 8. What You Should Know About Becoming an Excellent Subject 9. Techniques for Reaching the Somnambulistic State 10. A New Approach to Self-Hypnosis When All Else Fails 11. Psychological Aids and Their Function 12. The Nature of Hypnosis **120 Pages . . . $2**

The books listed above can be obtained from your book dealer or directly from Melvin Powers. When ordering, please remit 25c per book postage & handling.
Send 25c for our illustrated catalog of self-improvement books.

Melvin Powers

12015 Sherman Road, No. Hollywood, California 91605

NOTES

NOTES

NOTES